SABBATH MOOD HOMESCHOOL
PRESENTS

Living Science Study Guides

A Charlotte Mason Resource for Exploring
Science, a Vast and Joyous Realm

FORM 3-4 CHEMISTRY
(GRADES 7-9)

Accompanying the book *The Mystery of the Periodic Table* by Benjamin Wiker

FORM 3-4 CHEMISTRY

Second Edition: July 2021
ISBN: 9798533306836

www.SabbathMoodHomeschool.com

"But for the most part science as she is taught leaves us cold; the utility of scientific discoveries does not appeal to the best that is in us, though it makes a pretty urgent and general appeal to our lower avidities. But the fault is not in science—that mode of revelation which is granted to our generation, may we reverently say?—but in our presentation of it by means of facts and figures and demonstrations that mean no more to the general audience than the point demonstrated, never showing the wonder and magnificent reach of the law unfolded."

-CHARLOTTE MASON, *TOWARDS A PHILOSOPHY OF EDUCATION.*

Introduction

In this Form 3-4 study guide, students will learn about the development of modern chemistry, the formation of the periodic table, the structure and properties of matter, chemical reactions, and balancing chemical reactions.

Spine Text

This study guide accompanies the living book *The Mystery of the Periodic Table* by Benjamin Wiker. (Bethlehem Books, 2013).

- 163 pages, 18 chapters
- Reading Level: Grade 6 and up
- Prerequisites: none

Author Bio

When Benjamin Wiker (1960) was a university student he began reading "The Great Books". It was through studying great truths and great errors, the arguments of philosophers, theologians, and anti-theologians that he came to rediscover natural truth, goodness, and beauty. And more importantly, to know Jesus Christ. He has said, "When you are introduced to greatness you are not satisfied with anything less."

Benjamin Wiker went on to become, and still is, a professor. He has taught many courses on such interesting themes as political philosophy, history, Latin, mathematics, the history and philosophy of science, and many more.

He has authored/co-authored 13 books thus far, written and hosted a television show that is now the book, *Saints Vs Scoundrels*, but his greatest accomplishment is his marriage and seven children. (Jahncke, Michele)

Broken Links

If you notice a broken link in this study guide, please email the author for a replacement: Nicole@SabbathMoodHomeschool.com.

Schedule

This study guide includes 33 lessons, each requiring approximately 40 minutes. It can be scheduled in one of the following ways:

- Once a week for an entire year, allowing time for exams at the end of each term, and including other science subjects on the other days of the week, or
- Three times a week for 11 weeks allowing for exams during the 12th week.

Exams

You can download a digital copy of the exam questions for this study at https://qrs.ly/dgcq7r7.

Current Events

Each week, students should read a scientific current event. ScienceNewsforStudents.org is an excellent resource. Also, NewsELA, one of the best general news options for students, includes many scientific articles and allows you to change the difficulty of the text to your student's reading ability. You must create an account to view the articles on this site, but it is free and easy.

Science Notebook

Students should write narrations in their science notebook, including drawings, where appropriate, to better show what they have learned. This is not a test, so if they need to look at a diagram to copy it into their own book, that is acceptable. All reading, experiments, activities, and current events should be included, and each item should be dated. Students may also include quotes which they particularly liked from the reading. Learn more about keeping a science notebook in the SMH article, "Keeping a Science Notebook." (https://qrs.ly/racno3a)

Leisure Reading Suggestions

Your students may like to read more about this science topic during their free time, so choose a few of the following books to purchase or check out from your local library.

- *Mendeleyev and His Periodic Table* by Robin McKown (191 pp.)
- *Doctor Paracelsus* by Sidney Rosen (214 pp.)
- *Uncle Tungsten: Memories of a Chemical Boyhood* by Oliver Sacks (352 pp.)
- *Robert Boyle: Founder of Modern Chemistry* by Harry Sootin (142 pp.)
- *The Chemist Who Lost His Head, The Story of Antoine Laurent Lavoisier* by Vivian Grey (112 pp.)
- *Antoine Lavoisier: Scientist and Citizen* by Sarah R. Riedman (192 pp.)
- *The Invention of Air* by Steven Johnson (276 pp.)
- *The Radium Woman* by Eleanor Doorly (196 pp.)
- *Napoleon's Buttons: How 17 Molecules Changed History* by Penny Le Couteur (390 pp.)
- *Oxygen: The Molecule that Made the World* by Nick Lane (384 pp.)
- *Mauve: How One Man Invented a Color that Changed the World* by Simon Garfield (240 pp.)
- *Molecules of Murder: Criminal Molecules and Classic Cases* by John Emsley (252 pp.)
- *The Chemy Called Al* by Wendy Isdell (154 pp.)
- *The Periodic Kingdom: A Journey into the Land of the Chemical Elements* by P. W. Atkins (161 pp.)
- Find more biographies on the chemistry page at SabbathMoodHomeschool.com.

Supply List

If you are gathering supplies for the whole course, the following list should be helpful. If you will instead gather supplies for each lesson, see the Teacher Prep page for each week.

Find a digital list and links to suggested products at www.SabbathMoodHomeschool.com/Form3-supply-list.

HomeScienceTools.com or Supply House

Alka-seltzer tablets, 2-3
Ammonium nitrate. Alternatively, you can cut open a cold pack bag to remove the packet of little white balls. It's full of ammonium nitrate.
Beaker, 250 ml or a glass cup
Calcium chloride, granular, an Irritant OR Road Salt (Read the label to make sure it is pure calcium chloride; sometimes other chemicals are added to help melt snow.)
Connecting wires or alligator test leads
Dry yeast, 1 tablespoon or ¼ teaspoon potassium iodide Both work equally well. It just depends on whether you would like to use a common household item or a product that seems more *scientific*.
Graduated cylinder
LED diode, 1
Red litmus paper
Blue litmus paper
Safety glasses
Voltmeter or multimeter

Hardware Store or Hobby Store

Balloons, 2
Brick or wide-mouthed jar filled with water and tightly closed
Calligraphy pen and nib (crow quill nib and holder) or a toothpick (optional, see Lesson 18)
Candle. A votive candle would work best, but you can use another size as long as it is at least one inch shorter than your jar. (See Lesson 3)
Chopped up rose petals (optional)
Galvanized electrical box
Glow sticks, 3
Modeling clay, 2" chunk
Wooden dowel rods or skewers, 4

Grocery Store

Ammonia (optional, see lesson 18)
Aspirin
Beetroot, canned or fresh
Bendable drinking straw
Colorless soda (such as 7-Up)
Digital scale
Food coloring, any color
Fragrant spice, such as cloves or cinnamon, 1 tablespoon
Fruit juice, Cranberry, Pineapple, and 2 other kinds
Funnel
Hydrogen peroxide (6% solution), 50 ml (1.5 oz)
Iron supplement that contains ferrous sulfate to make your own ferrous sulfate solution OR purchase ferrous sulfate drops
Lemon juice
Liquid detergent
Liquid dish soap, 1 tablespoon
Matches
Milk of Magnesia
Plastic cups, 2 wide (9 oz) clear
Plastic cups, 6 small
Plastic deli-style condiment containers, 2
Plastic stirrer

Red cabbage
Rubber gloves
Rubbing alcohol (isopropyl alcohol)
Steel wool without soap
Straws
Turmeric
Water, distilled

Around the house
Baking soda
Burner to heat the pot of water
Cake pan
Coffee filter or paper towel
Drinking glass, tall
Glass bowl that fits over the saucepan (See Lesson 6)
Glass bowl, small, such as a cereal bowl, to act as a collecting dish
Glue
Hairdryer (optional, see lesson 18)
Ice cubes
Jar, small, such as a narrow jelly jar
Long-necked bottle (such as an empty, and clean, individual size soda bottle)
Masking tape
Measuring cups
Measuring spoons
Metal pie plate
Milk
Pan with a lid that is curved (see Lesson 6)
Paper
Paper towel
Pen
Pennies, 17
Piece of cardboard that will cover the top of the glass
Ruler
Salt
Saucepan, 3-quart, made of stainless steel, enamel or glass (not cast iron or aluminum)
Scissors
Sink or a large bowl
Steam iron (optional, see Lesson 18)
Tape
Tea (instant tea can be used)
Thermometer, candy or meat
Thick card stock
Toothpicks
Vinegar
Water
White Styrofoam egg carton with the top removed
Print Chemical Composition Worksheet (https://qrs.ly/1pcqhz9)

Print Acid Base Indicator Test table (https://qrs.ly/trcqibc)

Zipper baggies, large

LESSON 1

Have you ever watched or read a mystery a second time? The first time through, you are kept guessing and speculating on what happened. What is the truth? The author only unveils parts of the story, and you must pick up little clues to form a picture of what the true, complete story might be. Have you ever gotten to the end and realized you had it all wrong? Maybe some detail misled you, or you put too much emphasis on one aspect while overlooking another critical piece. However, if you watch the movie a second time, all the clues seem to fit into place. "*Why, of course!*" you might say, "*Why didn't I see that before?*"

As you settle in to read the history of chemistry, you will learn many foundational concepts that chemists use today. You will also learn why it is important to have imagination and why it is essential to question what you think you know. For example, when the ancient people held a chunk of metal in their hands, they only had a few clues to work with to uncover its mystery. Also, they only had a few tools to help them sort it out, and many times they didn't even know what questions to ask. We look back at their story and think, "*Why, of course! Why didn't they see that?*" But we must not judge too harshly. Instead, we might reconsider what we think we know today — are we making assumptions based on a few clues? We have better tools, but are we asking the right questions?

> *"The lessons of the ages have been duly set, and each age is concerned, not only with its own particular page, but with every preceding page. For who feels that he has mastered a book if he is familiar with only the last page of it?"*
>
> — CHARLOTTE MASON, SCHOOL EDUCATION, P. 160

Read: *The Mystery of the Periodic Table.* Chapter 1, "The Puzzle," pages 1-3. Place your bookmark at the beginning of chapter 2.

Please Note—Because you will be stopping partway through each of the chapters you read, I recommend that you find your stopping point and move your bookmark to that location before you begin reading.

Notebook: Write what you have learned in your science notebook. Please note that it is likely to take you as long, or nearly as long, to write your narration on each chapter as it takes you to read it. This chapter was relatively short, but subsequent chapters will be longer. If you find you cannot remember many of the ideas presented, then read smaller chunks before narrating. Take your time reading, understanding each paragraph before moving on.

Activity: Include a Periodic Table and a timeline in your science notebook.

As you read this book, the author will ask you to look up the elements mentioned so that you can become more familiar with their position. There is a Periodic Table inside the back cover of your book, which you can look at, but if you make a copy of it and attach it inside your science notebook, you can highlight each element as you read about it.

Supplies Needed—

- Photocopier
- Glue or tape
- Ruler

Procedure—

1. Photocopy the Periodic Table inside the back cover of your book. Adjust the size if needed to accommodate the size of your science notebook. There is also a smaller image displayed before chapter one.
2. Glue or tape it into your notebook.
3. On another page, prepare a timeline as follows:

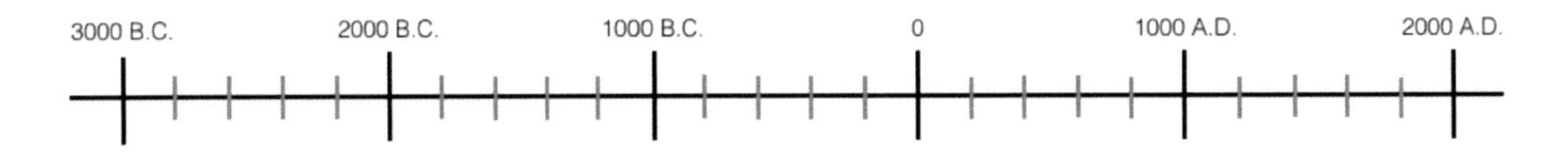

For Discussion: As you begin this study of chemistry, are you eager to learn? Or are you doubtful you will enjoy it? Either way, chemistry is everywhere in the world around you. So be observant as you go about your days. Watch for all the ways chemistry touches your life.

LESSON 2

Read: *The Mystery of the Periodic Table.* Chapter 2, "The First Chemists," pages 4-11. Place your bookmark at the beginning of chapter 3.

1st-century-BC Greek gold bracelets from a tomb near present-day Olbia, Ukraine.
Source: Walters Art Museum

Notebook: Write what you have learned in your science notebook. This section likely took you about 15 minutes to read. Therefore, it should take you approximately that same amount of time to write a narration. Please remember that your narration is the work of your education.

Add the Bronze and Iron age to your timeline.

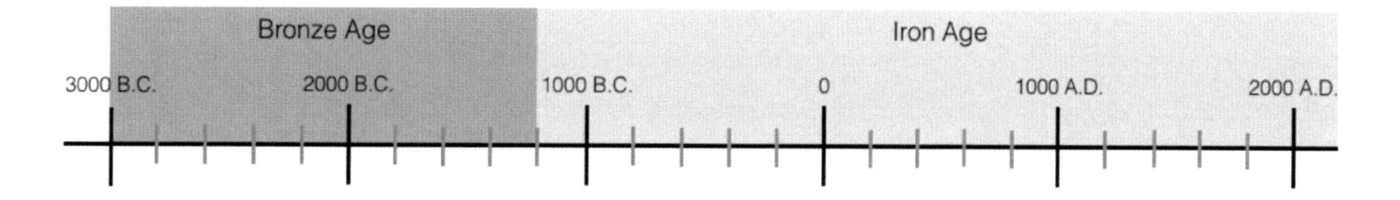

Use the Periodic Table that you put in your science notebook during the last lesson to look up the elements mentioned in this chapter. Mark them somehow, such as by coloring them in with a highlighter or colored pencil or by putting a checkmark in the box. If you color them, you could use a single color to note the following elements which were known to the ancients:

- Gold (Au) 79
- Silver (Ag) 47
- Copper (Cu) 29
- Lead (Pb) 82
- Tin (Sn) 50
- Iron (Fe) 26
- Mercury (Hg) 80
- Carbon (C) 6, in the form of diamonds and charcoal at that time in history
- Sulfur (S) 16, called brimstone at that time in history

Also mentioned in this chapter:

- Oxygen (O) 8

Activity: In this chapter, you read that the process of smelting removes elements from metal ore, so only pure metal remains. Observe this process by watching the video "Smelting Steel." (https://qrs.ly/kycqi9e, 6:52 min) There is little narration in this video, so remember what you learned on pages 9- 10: when iron ore is smelted, the element carbon is added as a reducing agent in the form of charcoal, and limestone is added as a flux. See if you can identify each of the materials used in the video.

LESSON 3—Experiment

Ancient scientists didn't have fancy tools to work with, nor did they have much interest in science just for the sake of it. Still, as you learned in the last two lessons, they recognized two valuable processes still used in the field of chemistry today – *separation* and *combination*. They used these processes to create tools and jewelry, but chemists today use the same techniques to make many products, from medicine to plastic. Even the salt on your table, if it does not specify that it is sea salt, is made by the process of combination.

The experiment you will try today will use a combination reaction.

Combination Reaction Definition: A combination reaction is a reaction where two reactants are combined into one product. Combination reactions are also known as synthesis because, through these reactions, new substances are synthesized.

Reactants are the starting materials in a chemical reaction. Reactants undergo a chemical change in which chemical bonds are broken, and new ones are formed to make products. In a chemical equation, reactants are listed on the left side of the arrow, while products are on the right side.

A familiar example of a synthesis reaction is the overall equation for photosynthesis:

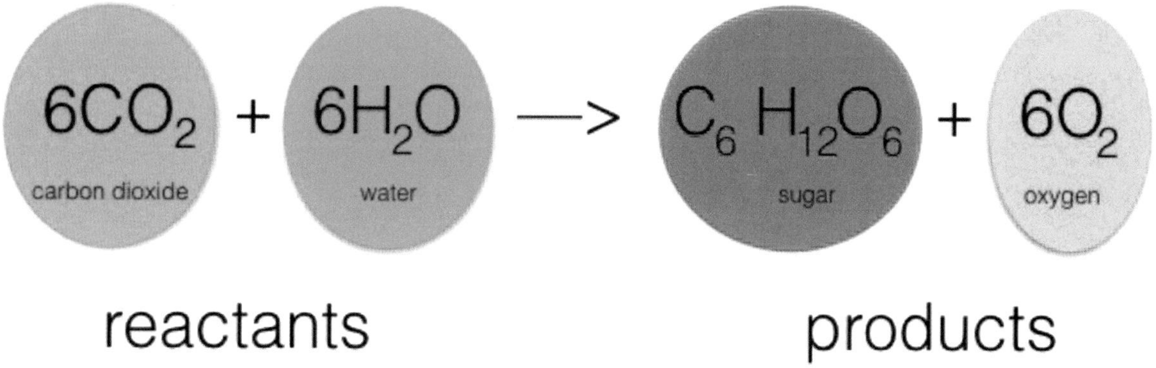

$$6CO_2 + 6H_2O \longrightarrow C_6H_{12}O_6 + 6O_2$$

carbon dioxide, water, sugar, oxygen

reactants — products

The reactants, carbon dioxide and water, undergo a chemical change in which chemical bonds are broken, and new ones are formed to make new products. In this case, glucose (sugar) and oxygen.

Activity: Testing synthesis reactions.

Supplies Needed—

- Candle. A votive candle would work best, but you can use another size as long as it is at least one inch shorter than your jar.
- Matches
- Metal pie plate
- Water
- Small jar, such as a narrow jelly jar
- Steel wool without soap
- Masking tape
- Vinegar

Please Note—A burning candle is an open flame that can cause burns. Liquid wax is hot and can cause burns to the skin. If you have long hair, tie it back and always be aware of where the flame is with respect to your clothing—do not lean over the flame.

Procedure—

PART 1

1. Light the candle and carefully drip a bit of melted wax in the middle of the pan.
2. Set the candle upright in the melted wax, making sure it is secure.
3. Fill the pan ⅔ full with water.
4. Put the jar upside down over the candle.
5. Observe what happens.
6. Note the height of the water inside the jar shortly after the candle goes out.

The water rising in the jar after the candle goes out replaces the substance in the air used up by the flame. What substance is that?

When a candle burns, it uses a lot of oxygen. When you limit the amount of oxygen, as you did by putting a lid over it, it runs out of materials.

We can represent the combustion of a candle like this:

$$2C_{18}H_{38} + 55O_2 \longrightarrow 36CO_2 + 38H_2O$$

The above equation is read this way: **two wax molecules** react with **55 oxygen molecules** to produce **36 carbon dioxide molecules** and **38 water molecules**.

PART 2

7. Remove the candle from the pie pan.
8. Dip your steel wool into vinegar and then stuff it into the bottom of the jar. Use enough so that it doesn't fall out when you turn the jar upside down. If necessary, use masking tape to stick it in place.
9. Place the jar with the steel wool upside down in the pan.
10. Let it sit for several days as you observe iron combine with oxygen to form iron oxide or rust.
11. Note the height of the water inside the jar.

This activity was adapted from Chemically Active by Vicki Cobb.

Notebook: Record the steps you took and what you learned from each experiment in your science notebook. Include drawings, if you would like.

For Discussion: You read that the ancients used chemistry to make tools and jewelry. But the Egyptians also used chemistry to make pottery, paint, makeup, and preserve mummies. But, of course, they didn't call it chemistry. They were meeting the needs and wants of their people.

One example was the creation of Egyptian blue, now known as calcium copper silicate, a synthetic blue pigment made up of a mixture of sand, limestone, bits of copper, and an alkali. They would have found all of these substances on the ground or the beds of dried-up salt lakes. The mixture was rolled into balls and baked in a furnace. Upon cooling, it became a glassy blue substance that was then ground down to make a blue-colored pigment.

This process sounds a little like an advanced recipe for mud cakes, which you probably made as a young child. Can you think of any ways you use chemistry without even thinking about it?

Optional Activity: If you would like to learn more about Egyptian blue and how art has facilitated the advancement of chemistry in the past, you can read the Chemistry World article "Egyptian blue: more than just a colour" (https://qrs.ly/macqiaf). You may also be interested in reading the book *Mauve: How One Man Invented a Color that Changed the World* by Simon Garfield or *Bright Earth: Art and the Invention of Color* by Philip Ball.

Egyptian blue ceramic ware, New Kingdom (1380-1300 BC)
Source: Walters Art Museum

LESSON 4

There is a debate going on in the field of education regarding whether students learn best by doing and experiencing or by learning the why behind the process. Do you think you learned more by doing the experiment in the last lesson or by reading about the idea of combination in the previous chapter of the text? Is it possible that you learned more by reading about it and experiencing it for yourself?

The scientists of ancient times, and even the middle ages, didn't always have the tools to experiment and see things for themselves, so often they had to use their imagination to come up with reasons why things are the way they are. In today's reading, you will learn about the theories of the men living in the 6th through the 1st century BC. They had no way to test their ideas, but they used their imagination. You might be surprised at how close to the truth some of them got.

Read: *The Mystery of the Periodic Table.* Chapter 3, "Earth, Air, Fire, Water," pages 12-18. Place your bookmark at the beginning of chapter 4.

Notebook: 1) Tell about the three ancient theories of earth's fundamental elements. Include drawings of the models where applicable. 2)Add the ancient philosophers and teachers mentioned in this chapter to your timeline.

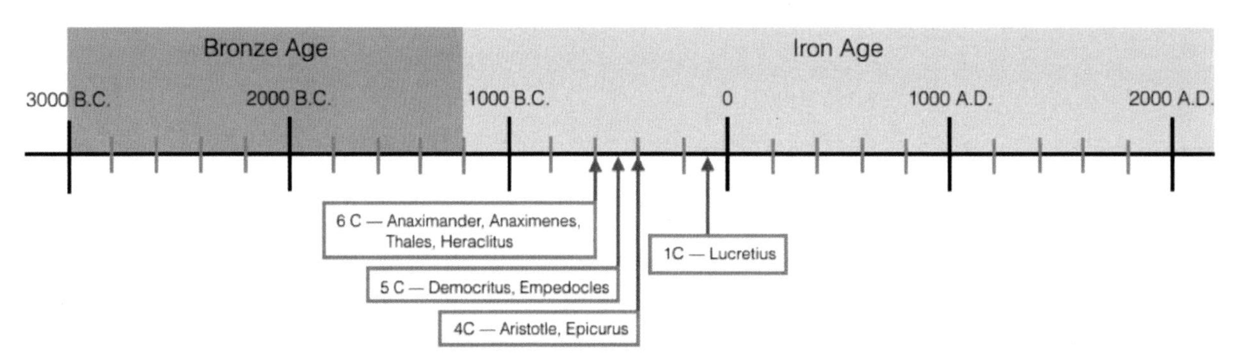

For Discussion: Do you think using your imagination in the field of science is important today? Do you think there might be things we still do not know about and do not have the instruments to test for yet?

LESSON 5

"The alchemists could not be accused of laziness or aversion to work in their laboratories. Paracelsus says of them: 'They are not given to idleness, nor go in a proud habit, or plush and velvet garments, often showing their rings on their fingers, or wearing swords with silver hilts by their sides, or fine and gay gloves on their hands; but diligently follow their labours, sweating whole days and nights by their furnaces. They do not spend their time abroad for recreation, but take delight in their laboratories. They put their fingers among coals, into clay and filth, not into gold rings. They are sooty and black, like smiths and miners, and do not pride themselves upon clean and beautiful faces.'

In these respects the chemist of to-day faithfully follows the practice of the alchemists who were his predecessors. You can nose a chemist in a crowd by the smell of the laboratory which hangs about him; you can pick him out by the stains on his hands and clothes. He also 'takes delight in his laboratory'; he does not always 'pride himself on a clean and beautiful face'; he 'sweats whole days and nights by his furnace.'"

— M. M. Pattison Muir, *The Story of Alchemy and the Beginnings of Chemistry*

Image Source: Joseph Wright of Derby, [Public Domain], via Wikimedia Commons

Read: *The Mystery of the Periodic Table.* Chapter 4, "The Alchemists," pages 19-27. Place your bookmark at the beginning of chapter 5.

Please Note—You will have an opportunity to try the experiments suggested in this chapter during the next lesson.

Notebook: 1) Tell about the spread of alchemy. 2) Then, recreate the chart on page 21 and explain the relationship between the known metals, the planets, and the days of the week.

Optional Activity: "*In August 2002 Prague was flooded by heavy rains. The results were catastrophic: billions of Czech korunas' worth of damage, seventeen dead, and forty thousand people evacuated from their homes. In the aftermath, the resilient residents dusted themselves off and started to pick up the pieces. The owner of house number 1, on Haštalská street was one such resident. A wall in his basement had collapsed. But the collapse wasn't all bad news. It revealed long-forgotten, walled up secret passages and subterranean alchemy laboratories...*" If you are interested, continue reading the article "Inside an ancient alchemy laboratory: Speculum Alchemiae" (https://qrs.ly/y3cqiak).

LESSON 6—Experiment

In the last chapter of the text, you learned that the alchemists forwarded the field of chemistry in three ways: by creating instruments, creating various compounds and recording the reaction of those compounds to heat, and discovering the usefulness of acids in breaking down compounds.

Today you will have an opportunity to try the two experiments mentioned in the last chapter: 1) the distillation of saltwater and 2) using the acid vinegar to separate the compound baking soda.

Activity: Distillation of saltwater and separation of baking soda.

Supplies Needed—

PART 1

- If following the procedure on page 24
    - 1 tablespoon of salt
    - 3 cups of water
    - A pan with a lid that is curved
    - A burner to heat the pot of water
- If following the procedure below
    - A 3-quart saucepan made of stainless steel, enamel or glass (not cast iron or aluminum)
    - A brick or wide-mouthed jar filled with water and tightly closed
    - A small glass bowl, such as a cereal bowl, to act as a collecting dish
    - 2 cups water + 1-2 cups more
    - 3-4 drops of food coloring, any color
    - 1 tablespoon of a fragrant spice, such as cloves or cinnamon
    - 1 tablespoon of salt
    - Glass bowl that fits over the saucepan
    - Ice cubes
    - A burner to heat the pot of water
    - Optional: chopped up rose petals

PART 2

- 6 tablespoons of vinegar
- 4 tablespoons of baking soda
- A tall drinking glass
- A long-necked bottle (such as an empty, and clean, individual size soda bottle)
- A balloon

Procedure—

PART 1—Distillation
Distillation was then, and is now, a fundamental process in chemistry and may account for 90% of all separation processes in the chemical industry.

Please Note—You may follow the instruction in the text on page 24 to perform this experiment, or you can follow the instructions below, which are a little more in-depth.

1. Set the saucepan on a stove burner
2. Put the brick or water-filled jar in the center of the saucepan.
3. Mix 2 cups of water with 3-4 drops of food coloring, 1 tablespoon salt, and 3 teaspoons of spice.
4. Pour this mixture into the saucepan.
5. Set the collecting dish on top of the brick or water-filled jar.
6. Set the glass bowl over the saucepan and fill it with ice and water.
7. Turn the burner to medium heat.
8. After the water in the saucepan has boiled for a few minutes, check to see if at least a tablespoon of water has collected in your collection bowl.
9. Turn off the burner.
10. CAREFULLY remove the collection bowl. Be cautious not to burn yourself on the steam or the hot pan or bowl. It is okay to let the apparatus sit long enough to cool completely.
11. Observe the liquid in the collection bowl — smell it, observe the color, and taste it.

This activity was adapted from *Chemically Active* by Vicki Cobb.

PART 2—Break down baking soda using the acid vinegar.

Please Note—If you do not have time to do this experiment today, you can save it for the next lesson.

The following instructions are from page 26 of the text.

1. Put three tablespoons of vinegar into a tall glass.
2. Then, over the sink or outside, add one teaspoon of baking soda to the glass.
3. Where did the bubbles come from? Was there air trapped in the vinegar? In the baking soda? In both?
4. Repeat the experiment with a small-necked bottle.
5. Put three tablespoons of vinegar into the bottle.
6. Add one tablespoon of baking soda to the bottle.
7. Then QUICKLY place the neck of the balloon over the neck of the bottle.

8. Now you know that air is being produced or released from the ingredients, which the alchemists missed.

Notebook: Record the steps you took and what you learned from each experiment in your science notebook. Include drawings, if you would like.

Why do you think you could use the simple distillation process to separate the water and salt in part 1, but the acid was needed to break apart the baking soda in part 2? Don't worry if you don't know the answer to this question yet. For now, you can write the question in your notebook. If you have any ideas, write those as well.

For Discussion: In Lesson 2, you learned about smelting, a process of separation. How does the separation process of distillation differ from smelting?

LESSON 7

In your last activity, you discovered that air was being produced or released from the ingredients you put in the bottle. The alchemists didn't consider the air any more than a fish considers water. Still, finally, a man named Johann Baptista van Helmont conducted an experiment similar to the one you did with the balloon in the last lesson. He was the first to trap air. But did he realize what he had found?

Johann Baptista van Helmont
Source: by Mary Beale, via Wikimedia Commons

Read: *The Mystery of the Periodic Table.* Chapter 5, ""This Spirit, Hitherto Unknown"," pages 28-33. Place your bookmark at the beginning of chapter 6.

Notebook: 1) Write what you have learned in your science notebook. 2) Add Johann Baptista van Helmont to your timeline.

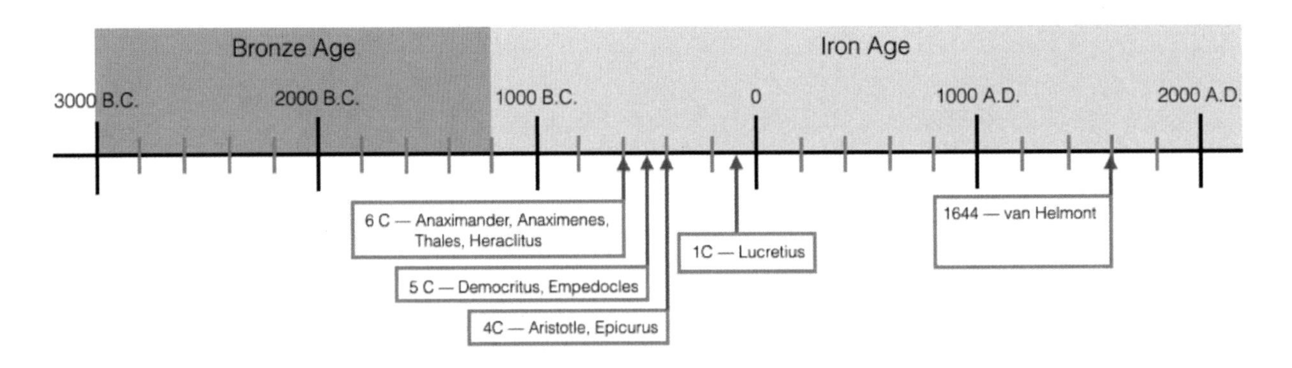

3) Look up the elements mentioned in this chapter. Mark or color the ones you haven't already noted on your Periodic Table.

1. Hydrogen (H) 1
2. Oxygen (O) 8
3. Nitrogen (N) 7
4. Carbon (C) 6

Activity: If you did not have time to complete both experiments during the last lesson, you could do those now.

For Discussion: "*One's logic is only as good as one's observations.*" (Benjamin Wiker, *Mystery of the Periodic Table*, p. 32) Have you ever looked for something but couldn't find it, only to realize it was right under your nose the whole time? That is how it was for many early scientists, but today our scientists have it all figured out, right? Or do you think even scientists today may miss a vital piece of information that is essentially right under their noses?

LESSON 8

Think about what you have learned so far through this study of chemistry. We use the term chemistry, but before the 17th century, we call what people did in this field alchemy, not chemistry. Early chemistry began with Johann Baptista van Helmont, whom you learned about in the last lesson. Some say his book *Ortus medicinae* aided the transition between alchemy and chemistry. It also significantly influenced Robert Boyle, who is said to have separated chemistry further from alchemy. Curiously, Boyle, not Helmont, is regarded as the first modern chemist and, therefore, one of the founders of modern chemistry.

Source: By Johann Kerseboom

Read: *The Mystery of the Periodic Table.* Chapter 6, "The Atomists Return," pages 34-41. Place your bookmark at the beginning of chapter 7.

Please Note—You will have an opportunity to try the experiments suggested in this chapter during the next lesson.

Notebook: 1) Write what you have learned about Robert Boyle's contributions to the field of chemistry in your science notebook. 2) Add Robert Boyle to your timeline.

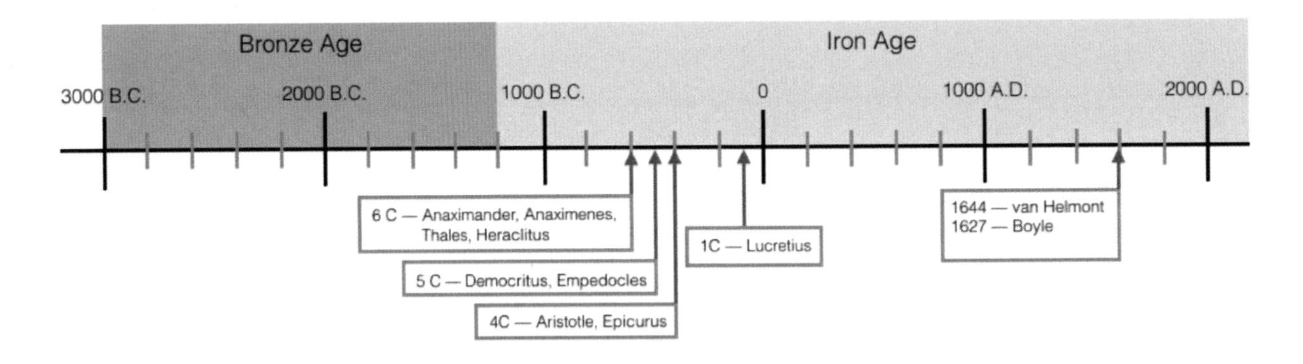

For Discussion: Modern chemistry didn't begin until the 17th century. Before that, we refer to alchemy, so why bother learning about it? Can you explain why these foundations were essential to kindling the field of chemistry?

LESSON 9—Experiment

Acid-base indicators are usually large molecules that react with acids and bases. When a reaction occurs, the structure of the molecules change, and so does the color. Today chemists make synthetic acid-base indicators, but natural indicators have been used since Robert Boyle discovered a way to test for acids and bases in the 17th Century. He described indicators extracted from roses and other plant materials in his book, *The Experimental History of Colors*. Still, they can also be made from many household items, such as fruit and vegetable juices, food colorings, inks, and even tea. Today you will have an opportunity to experiment with various indicators to observe the color change for yourself.

Activity: Testing acids and bases with various indicators.

Supplies Needed—

- Rubbing alcohol (isopropyl alcohol)
- Turmeric
- Red and blue litmus paper
- Distilled water
- Plastic stirrer
- White Styrofoam egg carton with the top removed
- Straws
- Toothpicks
- Red cabbage
- Beetroot, canned or fresh
- Teaspoon measuring spoon
- Plastic cups
- Masking tape
- Pen
- Aspirin
- Baking soda
- Milk of Magnesia (tablets or liquid)
- Salt
- Liquid detergent
- Colorless soda (such as 7-Up)
- Vinegar
- Lemon juice
- Tap water

Procedure—

1. Print out the Acid Base Indicator Test table (https://qrs.ly/trcqibc) to record your findings, or prepare a table in your science notebook.
2. Prepare your indicators and label each cup.
    a. Turmeric indicator – Add 1 teaspoon of turmeric to a ¼ cup of rubbing alcohol. Stir and allow to stand for several minutes before using.
    b. Red cabbage solution – Chop 1 cup of red cabbage. Add it to a cup of water and heat the mixture on the stove or in a microwave until the water is a deep purple color. Separate the solids from the liquid, allow the liquid to cool, then pour it into a plastic cup.

c. Beetroot indicator — Chop 1 cup of fresh beetroot. Add it to a cup of water and heat the mixture on the stove or microwave until the water is a deep purple color. Separate the solids from the liquid, allow the liquid to cool, then pour it into a plastic cup.

3. Prepare your test solutions and label each cup.
    a. Aspirin solution — 3 tablets in a cup of distilled water.
    b. Baking soda solution — 1 teaspoon of baking soda in a cup of distilled water.
    c. Milk of Magnesia solution — 1 tablet or 2 tablespoons in a cup of distilled water.
    d. Salt solution — 1 teaspoon in a cup of distilled water.
    e. Detergent solution — 1 teaspoon liquid detergent in a cup of distilled water.
    f. Colorless soda
    g. Vinegar
    h. Lemon juice
    i. Tap water
4. Test each solution with red litmus paper:
    a. Transfer a drop of one solution onto the red litmus paper with a stirrer.
    b. Record the color that is produced in Table 1.
    c. Wash off the stirrer with distilled water.
    d. Repeat the test with the other solutions and record your observations.
    e. Please note that you can use a piece of litmus paper to test two solutions using each end.
5. Repeat step 4 using blue litmus paper.
6. Test each solution with a turmeric indicator:
    a. Place a straw in a solution to be tested. Put your finger over the top of the straw. Remove the straw from the solution, and by slightly raising your finger, transfer 1 drop of the solution to one of the wells in the egg carton. Transfer 1 drop of the turmeric indicator solution to the same well and mix with a toothpick.
    b. Observe the color, and record your observations in Table 2. Is the solution acidic or basic?
    c. Repeat the test with the other solutions and record your observations.
    d. Rinse the egg carton with distilled water.
7. Test each solution with a red cabbage indicator:
    a. Transfer 2 teaspoonfuls of the solution to be tested to one of the wells in the egg carton.
    b. Transfer 1 teaspoonful of the red cabbage indicator solution to this well and mix with a plastic stirrer or straw. Observe the color, and record your

observations in Table 4. Determine if the solution is acidic, basic, or neutral and whether it is weak or strong.
   c. Repeat the test with the other solutions and record your observations.
8. Test each solution with the beetroot juice indicator:
   a. Place a straw in a solution to be tested. Put your finger over the top of the straw. Remove the straw from the solution, and by slightly raising your finger, transfer 1 drop of the solution to one of the wells in the egg carton. Transfer 1 drop of the beetroot juice indicator solution to the same well and mix with a toothpick.
   b. Observe the color, and record your observations in Table 3. Is the solution acidic or basic?
   c. Repeat the test with the other solutions and record your observations.
   d. Rinse the egg carton with distilled water.

Notebook:

- What color is litmus under acidic conditions? What color is litmus under basic conditions?
- Is the red cabbage indicator more useful than an indicator such as litmus for determining the acidity or basicity of a solution?
- If a solution turns the red cabbage indicator green, is the solution acidic or is it basic?
- If a solution turns the red cabbage indicator violet, is the solution acidic or is it basic?
- If a solution turns the red cabbage indicator bright yellow, is it safe to touch or taste?
- What color is turmeric under acidic conditions?
- What color is turmeric under basic conditions?
- What color is beetroot juice under acidic conditions?
- What color is beetroot juice under basic conditions?

This activity was adapted from Chemistry Experiments for the Home.

LESSON 10

The sample timelines included in this study guide show the year of birth of the person included because it is interesting to look at the age of those who made advances in the field of chemistry. You will find that much progress was made by people who were somewhat close in age. They were contemporaries, which means a person of roughly the same age as another, if not friends.

Scientific discoveries were being made in all fields across Europe during the late 17th century. You've read about the discoveries made by van Helmont and Boyle. Harvey discovered the circulation of blood, Roemer measured the speed of light, Torricelli detected atmospheric pressure, and Leeuwenhoek built a microscope and looked at everything from bacteria in rainwater to red blood cells. Brandt discovered glowing phosphorus, and Newton proposed his theory of gravitation. It was a fascinating time in the history of all scientific fields. However, today you will learn about a wildly imaginative man who practically stopped chemistry's progress for nearly 100 years by introducing an error that derailed many chemists.

Johann Becher
Source: by Gandvik, [Public Domain], via Wikimedia Commons

Read: *The Mystery of the Periodic Table.* Chapter 7, "The Strange Tale of Phlogiston," pages 42-47. Place your bookmark at the beginning of chapter 8.

Notebook: 1) Write what you have learned in your science notebook. 2) Add Johann Becher to your timeline.

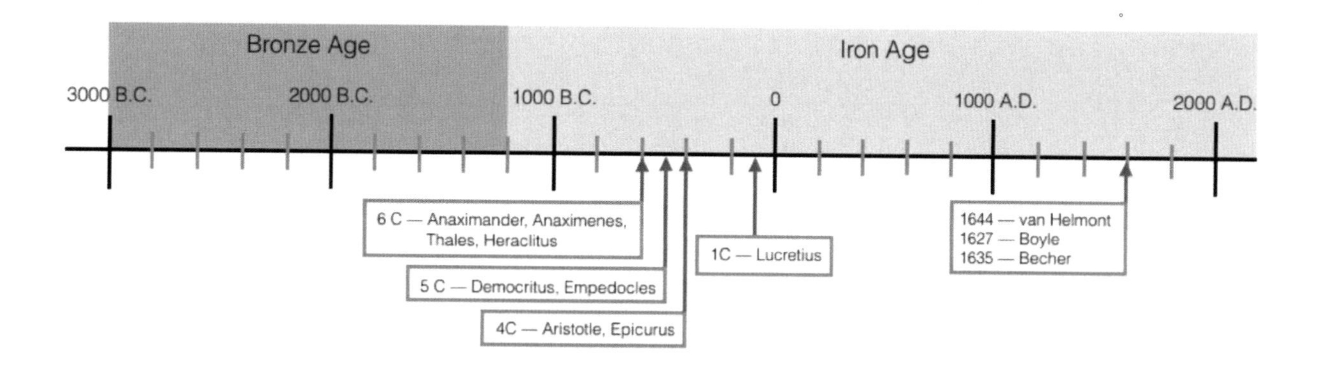

3) Look up the elements mentioned in this chapter. Mark or color the ones you haven't already noted on your Periodic Table.

1. Hydrogen (H) 1
2. Oxygen (O) 8
3. Zinc (Z) 30
4. Carbon (C) 6

For Discussion: In the text, you read: "*Chemists believed in phlogiston for over a hundred years, and it was difficult to get them to change their minds. Some never did.*" Today scientists have access to microscopes so powerful that they can see the structure of an atom. They don't have to guess what could be going on at the microscopic level. But do you think they may still overlook things? Do you think it is still important for them to use their imagination?

Additional Resource: If you are interested in learning more about Johann (John) Becher, I encourage you to spend some of your free time reading chapter 3 of the book *Crucibles* by Bernard Jaffe.

LESSON 11

In the last chapter of the text, you learned about the two-edged contribution of Johann Becher. Can you recall the story? As you begin the next chapter, you will notice that 100 years have passed.

Joseph Priestley, the next chemist the text will introduce to you, began his scientific study by attending chemistry lectures. However, his interest was soon redirected by his new acquaintance, Benjamin Franklin. Franklin interested Priestley in electricity, and he pursued it much the way you are beginning your study of chemistry — he studied its history and tried the original experiments. "*I was led in the course of my writing [an electrical] history, to endeavor to ascertain several facts which were disputed, and this led me by degrees into a larger field of original experiments in which I spared no expense that I could possibly furnish.*"[1] Today, you will read about Priestley's significant contribution to the field of chemistry, thanks to his continued interest in scientific experimentation.

Joseph Priestley

Source: by Ellen Sharples, [Public Domain], via Wikimedia Commons

Read: *The Mystery of the Periodic Table.* Chapter 8, "Mr. Priestley Clears Things Up," pages 48-53. Place your bookmark at the beginning of chapter 9.

---

[1] Jaffe, Bernard. *Crucibles the Story of Chemistry from Ancient Alchemy to Nuclear Fission.* New York: Dover Publications, 1976. 41. Print.

Please Note—You will have an opportunity to try the experiments suggested in this chapter during the next lesson.

Notebook: 1) Give an account of how Priestley discovered oxygen and how he explained what he had found. 2) Add Joseph Priestley to your timeline.

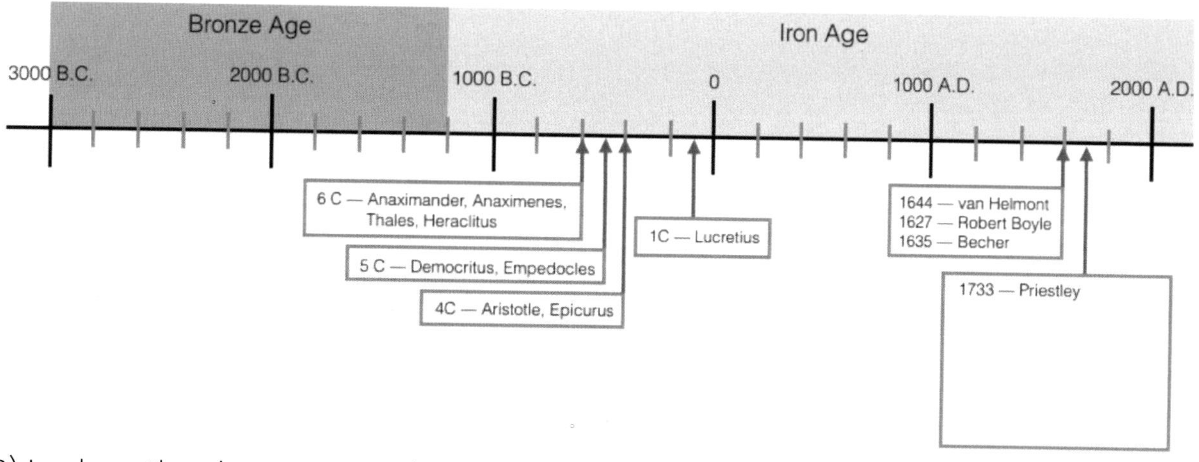

3) Look up the elements mentioned in this chapter. Mark or color the ones you haven't already noted on your Periodic Table.

1. Oxygen (O) 8
2. Mercury (Hg) 80
3. Zinc (Z) 30

Additional Resource: If you are interested in learning more about Joseph Priestley, I encourage you to spend some of your free time reading chapter 4 of the book *Crucibles* by Bernard Jaffe.

For Discussion: Joseph Priestley said: "*The more elaborate our means of communication, the less we communicate.*" Do you think he was correct? What would he think about our ways of communicating today?

LESSON 12—Experiment

Today you will have an opportunity to make a pneumatic trough as outlined in the last chapter.

Activity: Make and test a pneumatic trough.

Supplies Needed—

- Sink or a large bowl
- Water
- Drinking glass
- Piece of cardboard that will cover the top of the glass
- Bendable drinking straw

Procedure—

The following instructions are from pages 49-50 of the text.

1. Fill a sink or a large bowl with about 3-4 inches of water.
2. Fill a drinking glass to the very brim with water.
3. Place a piece of cardboard over the glass.
4. Holding the cardboard tightly over the glass, quickly (but carefully) turn the glass upside down and submerge the rim of the glass about a half-inch into the water.
5. Remove the cardboard.
6. Bend your drinking straw and slip the short end under the upside-down glass. (The image on page 50 shows the "straw" going all the way to the top of the glass, but that is not necessary, as the air bubbles will rise.)
7. Blow air into the straw.
8. Note that the exhaled gas displaces the water in the glass.

Notebook: Record the steps you took and what you learned from this experiment in your science notebook. Include drawings, if you would like.

LESSON 13

Are you a shy person, or are you pretty outgoing? Today you will read about a scientist named Henry Cavendish who was extremely shy. Long after his death, James Clerk Maxwell (the father of modern physics) went through Cavendish's papers and found that he had discovered as many as five laws of science that were attributed to others because Cavendish never shared his findings with anyone. He did, however, share the experiment you will read about today with the Royal Society of England.

Henry Cavendish
Source: Charles Turner

Read: *The Mystery of the Periodic Table.* Chapter 9, "Mr. Cavendish and Inflammable Air," pages 54-58. Place your bookmark at the beginning of chapter 10.

Notebook: 1) Write what you have learned in your science notebook. 2) Include the chemical equation for one of the acid - metal reactions talked about on page 56 and then write an explanation of that reaction. 3) Add Henry Cavendish to your timeline.

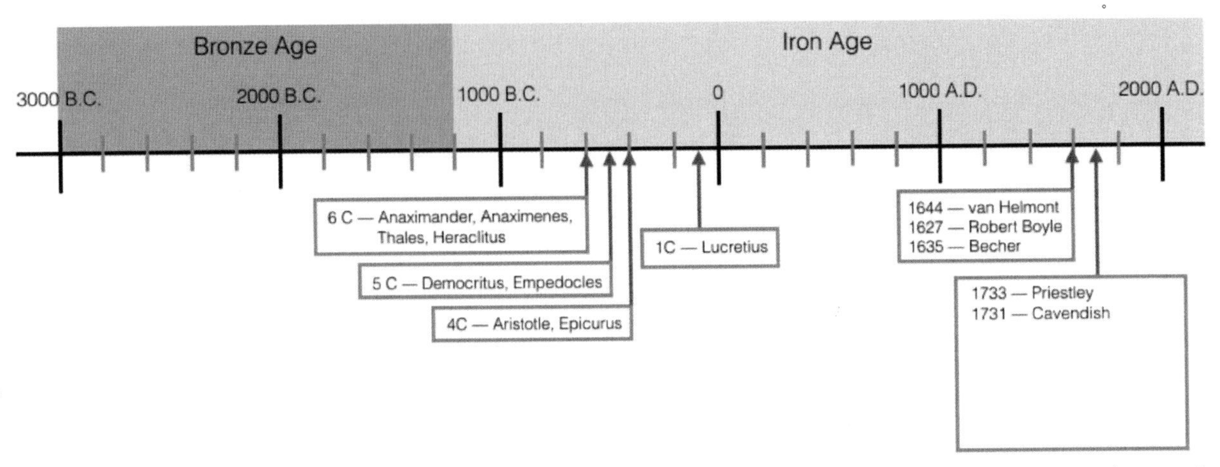

4) Look up the elements mentioned in this chapter. Mark or color the ones you haven't already noted on your Periodic Table.

1. Hydrogen (H) 1
2. Oxygen (O) 8
3. Iron (Fe) 26
4. Sulfur (S) 16
5. Chlorine (Cl) 17
6. Zinc (Z) 30
7. Tin (Sn) 50

Activity: To observe the hydrogen-oxygen reaction that makes water, watch the video "How to make water -- Hydrogen Oxygen reaction" (https://qrs.ly/4scqiex, 2:11 min). The assignment is to watch the video for demonstration purposes only; however, you will notice that it is an instructional video. Please do not attempt this experiment on your own without the permission and direct supervision of an adult.

Additional Resource: If you are interested in learning more about Henry Cavendish, I encourage you to spend some of your free time reading chapter 5 of the book *Crucibles* by Bernard Jaffe.

For Discussion: Do you think it makes the discoveries of Priestley and Cavendish any less important because they didn't know what they had discovered? They had some details right and some wrong, yet they are included in every book relating to the history of chemistry. Some of the books you will use to study science will have incorrect details. That may be clear to you as you read along, or maybe you won't know the author is wrong as you read. Do you think you can still learn something from a person who is partially wrong about a subject?

LESSON 14

You may have studied the famous artist Jacques-Louis David, who painted a double portrait of Antoine Lavoisier and his wife, Marie-Anne. Lavoisier was a chemist and a powerful man in France before the revolution. His wife is said to have been an intelligent, cultured woman with a passion for chemistry that matched her husband's. She took drawing lessons from the famous David and put those skills to good use by illustrating her husband's book *Elementary Treatise of Chemistry*.

Take a few minutes to study the painting below. Notice all of the details and what they tell you about the life of this remarkable couple.

Read: *The Mystery of the Periodic Table.* Chapter 10, "Chemistry's French Revolution," pages 59-67. Place your bookmark at the beginning of chapter 11.

Notebook: 1) Write what you have learned in your science notebook. 2) Include the Law of Conservation of Mass and an explanation of what it means. 3) Add Antoine Lavoisier to your timeline.

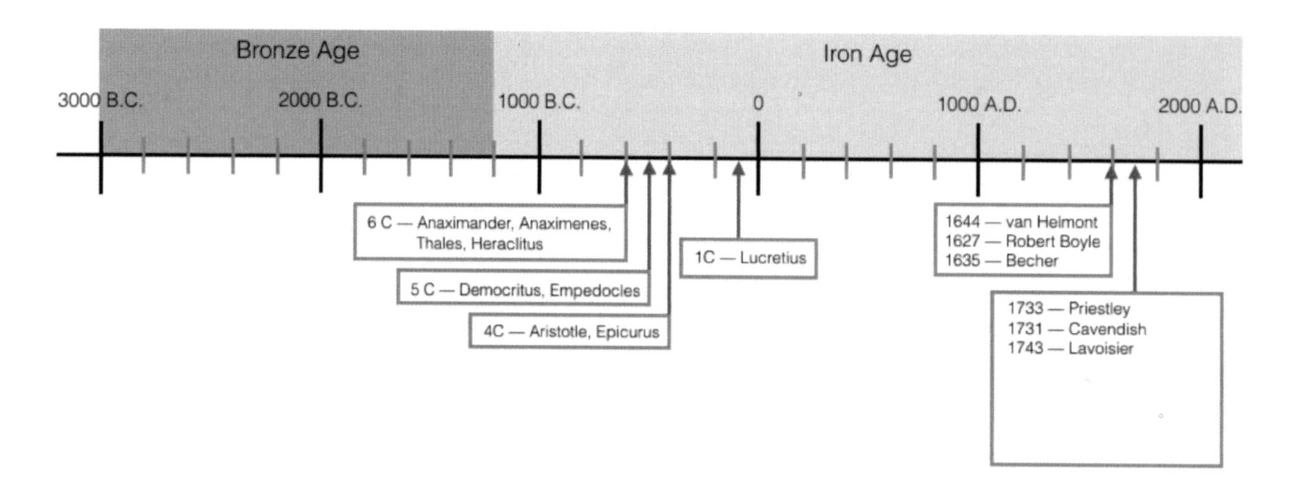

4) Look up the elements mentioned in this chapter. Mark or color the ones you haven't already noted on your Periodic Table.

1. Mercury (Hg) 80
2. Oxygen (O) 8
3. Hydrogen (H) 1
4. Nitrogen (N) 7

Additional Resource: If you are interested in learning more about Antoine Lavoisier, I encourage you to read one of the following books during your free time: *The Chemist Who Lost His Head* by Vivian Grey, or *Antoine Lavoisier* by Sarah R. Riedman, or chapter 6 of *Crucibles* by Bernard Jaffe.

For Discussion: On page 60 of the text, you read: "*and once Lavoisier decided something, it was already half-way to being accomplished.*" How do you think he earned that reputation?

LESSON 15—Experiment

In the last lesson, you read: "*Lavoisier was the first to make clear a very important principle in chemistry, one which is used in the [mentioned] experiments. It has come to be called the 'conservation of mass.'*" (p. 66-67) The Law of Conservation of Mass (or Matter) in a chemical reaction can be stated: In a chemical reaction, matter is neither created nor destroyed.

Today you will try the baking soda and vinegar reaction again, but this time you will pay attention to another aspect of the reaction.

Activity: Prove the conservation of mass.

Supplies Needed—

- 2 small plastic cups
- 6 tablespoons of vinegar
- 4 tablespoons of baking soda
- 2 large zipper baggies
- Digital scale
- 2-3 Alka-Seltzer tablets
- A tall drinking glass
- A long-necked bottle (such as an empty, and clean, individual size soda bottle)
- A balloon (in case you would like to use it in part 2)

Procedure—

PART 1

1. Prepare a data table in your notebook that includes the following fields:

Initial Mass (g)	Final Mass (g)	Change in Mass (g)

2. Fill one cup halfway with vinegar.
3. Fill a second cup halfway with baking soda.
4. Carefully put both cups in the plastic bag. Do not spill the contents of either cup.
5. Determine the mass of the cups and their contents in the plastic bag by weighing them on a digital scale. Write the values in your data table.
6. Seal the plastic bag.

7. Without opening the bag, pour the vinegar into the cup of baking soda.
8. Without opening the bag, record the mass of the contents of the plastic bag. Take care not to break the seal of the plastic bag.

Notebook: Describe what happened when you poured the vinegar into the cup of baking soda. How does this experiment demonstrate the Law of Conservation of Mass? If your change in mass was not zero, can you explain why that might be?

PART 2

Design an experiment to demonstrate the law of conservation of mass using an Alka-Seltzer tablet. Be specific in your procedures so another person or group could reproduce your investigation and gather the same data and observations. Once the procedures are written, conduct your investigation. Be sure to collect data and observations during each trial. If you change your procedures along the way, make adjustments to your procedures on paper. When you are through, write up your experiment so that another person can conduct it without you there. Include a supply list, procedure instructions, what results you expect, and an explanation of those results.

Optional — if you are working in a group and have time, you could have groups swap their experiments and follow the procedures as written.

This activity was adapted from Law of Conservation of Matter Lab.

Notebook: What did you learn from these activities?

For Discussion: In chapter 8 of your text, you read: "*Now one day—and we know the day, August 1, 1774—Priestley put calx of mercury underneath a glass. He focused the sun's hot rays on the calx with his new 12" diameter magnifying glass. It began to give off a gas. The calx of mercury changed back into mercury, and Priestley trapped the gas with his pneumatic trough. ... Without really thinking about it Priestley exposed the candle to the gas. The flame suddenly flared into brilliance. What was this wondrous gas?*" (p. 51) Priestly had isolated oxygen, but he called it dephlogisticated air.

Three years later, he met Antoine Lavoisier, who carefully listened as Priestley shared the details of his experiments. When they parted, Lavoisier obtained a sample of the red mercury powder Priestley had used and repeated his experiments. Thus, by following Priestley's lead, Lavoisier was finally able to discredit the 100-year-old theory of phlogiston. What do you think would have happened if Priestley had not shared his results or if he had shared them in a sloppy, unclear way?

LESSON 16

On February 2, 1790, Antoine Lavoisier wrote a letter to Benjamin Franklin stating: "*Here, then: a revolution [in science and chemistry] has taken place in an important part of human knowledge since your departure from Europe ... I will consider this revolution to be well advanced and even completely accomplished if you range yourself with us. ... After having brought you up to date on what is happening in chemistry, it would be well to speak to you about our political revolution. We regard it as done and without any possibility of return to the old order.*"[2] Today, you will read a little about both the revolution in chemistry that Lavoisier mentioned and the revolution in France that, four years later, would cut short the life of the man who had done so much for both the scientific community and the people of his country.

Read: *The Mystery of the Periodic Table*. Chapter 11, "A Revolution in Names," pages 68-72. Place your bookmark at the beginning of chapter 12.

Notebook: 1) Write what you have learned in your science notebook. 2) Look up the elements mentioned in this chapter. Mark or color the ones you haven't already noted on your Periodic Table.

Non-metals known in Lavoisier's time:

- Sulfur (S) 16
- Phosphorus (P) 15
- Carbon (C) 6
- Oxygen (O) 8
- Nitrogen (N) 7, called Azote at that time in history
- Hydrogen (H) 1

Other metals mentioned in this chapter, but not identified by the people of this time:

- Calcium (Ca) 20
- Barium (Ba) 56

Metals known in Lavoisier's time:

- Antimony (Sb) 51
- Arsenic (As) 33
- Bismuth (Bi) 83
- Cobalt (Co) 27
- Copper (Cu) 29
- Gold (Au) 79
- Iron (Fe) 26
- Lead (Pb) 82
- Manganese (Mn) 25
- Mercury (Hg) 80
- Molybdenum (Mo) 42
- Nickel (Ni) 28
- Platinum (Pt) 78
- Silver (Ag) 47
- Tin (Sn) 50
- Tungsten (W) 74
- Zinc (Z) 30
- Magnesium (Mg) 12

[2] Cohen, I. B. *Revolution in Science*. Cambridge: Belknap of Harvard UP, 1985. Print.

Please Note—The above list is long, but you should be able to quickly look up many of the elements listed since you have looked them up before, and your periodic table will be more than a quarter filled in when you are through.

For Discussion: Antoine Lavoisier is quoted as having said: "*The human mind adjusts itself to a certain point of view, and those who have regarded nature from one angle, during a portion of their life, can adopt new ideas only with difficulty.*" Do you think he is correct? How can a person guard themselves against getting set in their opinions?

LESSON 17

Today you will read about the discovery of a scientific "*law which, for the first time, made chemistry a mathematical science.*"[3] As with most discoveries, there were several people seeking answers to the same question. Some got the answer right and some wrong, but by communicating their ideas with the scientific community, eventually, the truth was found.

Read: *The Mystery of the Periodic Table.* Chapter 12, "Nature Never Creates Other Than Balance in Hand," pages 73-80. Place your bookmark at the beginning of chapter 13.

Notebook: 1) Write what you have learned in your science notebook. 2) Include the Law of Definite Proportions and an explanation of what it means. 3) Add Joseph Proust to your timeline.

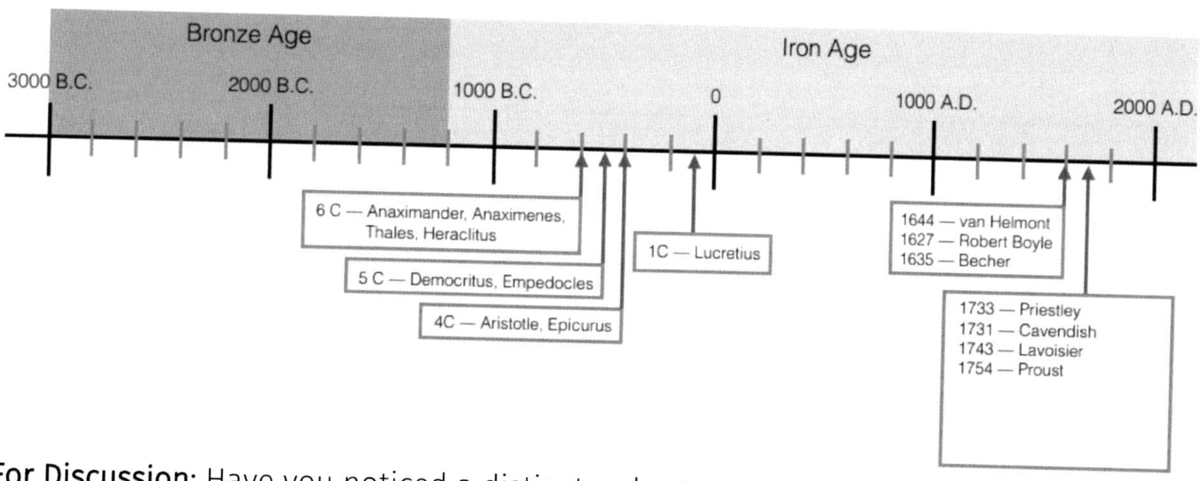

For Discussion: Have you noticed a distinct order to nature, such as the growth rings in a tree trunk, the water cycle, the seeds on a strawberry, snowflakes, honeycomb, the movement of the planets, the petals on a flower, and a spider web? Have you considered that even things that seem random, such as the composition of dirt, may have some order?

Joseph Proust said, "*The stones and soil beneath our feet, and the ponderable mountains, are not mere confused masses of matter; they are pervaded through their innermost constitution by the harmony of numbers.*"

---

[3] Jaffe, Bernard. *Crucibles the Story of Chemistry from Ancient Alchemy to Nuclear Fission.* New York: Dover Publications, 1976. 92. Print.

LESSON 18—Experiment

In Lesson 6, you performed two experiments. In the first one, you mixed the water and the salt as you would mix sand and salt, and then you separated the two again through distillation. Figuring out a way to separate the sand and salt would be an interesting experiment too.

In a separate experiment, you could have frozen the water, crushed the ice, boiled it to steaming water vapor, and yet, all the time, you would still be working with the pure substance water. Each of those examples are physical changes. They do not change the chemistry of water.

What you demonstrated in the second experiment, however, was not a simple physical change. Baking soda and vinegar are pure substances, just like water, but when you combined the two, the molecules of both substances underwent a chemical change.

Chemical changes happen through events called chemical reactions. In a chemical reaction, the bonds between atoms in the reactants are broken, the atoms rearranged, and new bonds between the atoms are formed to make a different product.

You also observed a chemical reaction when the candle burned in lesson three. The wax reacted with oxygen in the air to produce carbon dioxide and water. You can look back at the chemical equation shown on page 12 to remember how the atoms rearranged in that chemical reaction.

Today a scientist can confirm that a chemical change has taken place by performing a chemical analysis of the products. However, the scientists you have been studying hadn't even proved the existence of atoms yet. They could only tell if a chemical reaction had occurred by their observations.

Easily observed physical effects usually accompany a chemical reaction. For instance, you might notice a color change or the formation of gas. You might also record either an increase or decrease in temperature, and lastly, you might observe the formation of a precipitate. A precipitate is a solid formed in a chemical reaction that is different from either of the reactants.

Activity: Observe the formation of a precipitate.

Supplies Needed—

- 2 cups of cooled strong tea (instant tea can be used)
- Iron supplement that contains ferrous sulfate, to make your own ferrous sulfate solution OR ferrous sulfate drops
- Liquid measuring cup
- Water, room temperature and warm
- Coffee filter or paper towel
- Funnel
- 6 small plastic
- Cranberry juice
- Pineapple juice
- 2 other varieties of fruit juice
- Steel wool without soap
- Vinegar
- Optional activity:
    - Paper
    - Calligraphy pen and nib (crow quill nib and holder) or a toothpick
    - Hairdryer (optional)
    - Ammonia
    - Lemon juice, milk, and/or baking soda solution (1 teaspoon baking soda mixed with ¼ cup water)
    - Candle or steam iron

Procedure—

CAUTION: DO NOT DRINK ANY OF THE SOLUTIONS YOU PREPARE IN THIS ACTIVITY.

1. Prepare two cups of strong tea, and let it cool.
2. To prepare a ferrous sulfate solution
    a. Combine 5 iron tablets with a ½ cup of water in a measuring cup. Let the tablets soak, stirring occasionally, until the coating has completely dissolved.
    b. Remove the tablets from the water and rinse under a faucet. They should be white.
    c. Rinse out the measuring cup, fill it with about ¾ cup warm water, and put the tablets in the cup.
    d. Stir until the tablets break up.
    e. Set a coffee filter in a funnel and strain the iron water solution through the filter.

f. The *filtrate* will pass through the coffee filter, and the filter will trap the *residue*

3. Pour ¼ cup of cooled tea into each of the small plastic cups.
4. Add the ferrous sulfate solution to a cup of tea until a black precipitate forms.

   The black precipitate that was formed is known as iron gall ink. The iron salts in the ferrous sulfate solution reacted with the tannic acids in the tea to form this precipitate which was used for hundreds of years as the standard writing and drawing ink in Europe. Save this solution for the optional activity below.

5. Test one of the fruit juices by adding two tablespoons of juice to a cup of tea.
6. Repeat with each type of juice.

   Which fruit juices show the iron tannate precipitate when added to the tea? If you wish to make a fruit-flavored tea that remains clear, which fruit juice should you use?

7. Put some steel wool into a cup of tea.

   Does a precipitate form? Where?

8. If the steel wool looks a little rusty when sitting in the tea, add some vinegar and see if you get iron tannate to form. The formation of rust means that the iron has combined with oxygen instead of with tannic acid. Vinegar prevents the iron from combining with oxygen and frees it to combine with tannic acid.

Optional Activity: Write with your iron gall ink.

1. Dip a pen nib or toothpick in the iron gall ink and write or draw on a piece of paper. Notice if the ink changes color as it is exposed to air.
2. Let the ink dry. You can use a hairdryer to help it dry faster.
3. Hold the dried writing over some ammonia fumes, which escape from an open ammonia bottle. Notice if the ink changes color.

   In the presence of the ammonia fumes, the iron reacts with oxygen in the air to form iron oxide or rust. When the ammonia is removed, the iron reacts with the tannic acid to form iron tannate again. This is called a *reversible reaction*.

4. Remove the paper from the ammonia. Notice if the ink changes color again.
5. Dip a clean pen nib or toothpick in lemon juice, milk, or baking soda solution and write or draw on a piece of paper.

6. **WITH THE HELP OF AN ADULT** hold the paper about two inches over a candle flame and pass it back and forth. Be careful that the paper does not ignite. Alternatively, you can pass a hot iron over it. Notice if the "invisible ink" changes color.

   Heat causes other materials in the "invisible inks" to react with the paper and the oxygen in the air, leaving behind the carbon, which appears dark brown or black.

This activity was adapted from *Chemically Active* by Vicki Cobb.

Notebook: 1) Record the steps you took and what you learned from this experiment in your science notebook. Include drawings, if you would like. 2) List the physical indicators that usually accompany a chemical reaction. 3) Finally, think about each of the experiments you have performed during your chemistry lessons and note whether you saw one of these effects.

Old Man in a Turban by Rembrandt, circa 1638, pen and iron-gall ink on paper
Source: National Gallery of Victoria, Melbourne, [Public Domain], via Wikimedia Commons

LESSON 19

On your timeline, you noted Democritus (5C), Epicurus (4C), and Lucretius (1C), who were the first atomists. Robert Boyle (17C) was also an atomist, but it was not until John Dalton came along that a sound theory of the atom was proposed. With few exceptions, Dalton's atomic theory remains valid in modern chemistry.

John Dalton
Source: Charles Turner

Read: *The Mystery of the Periodic Table.* Chapter 13, "Mr. Dalton and His Atoms," pages 81-87, put your bookmark after the image on page 87.

Notebook: 1) Write what you have learned in your science notebook. Include the Law of Multiple Proportions and an explanation of what it means. 2) Add John Dalton to your timeline.

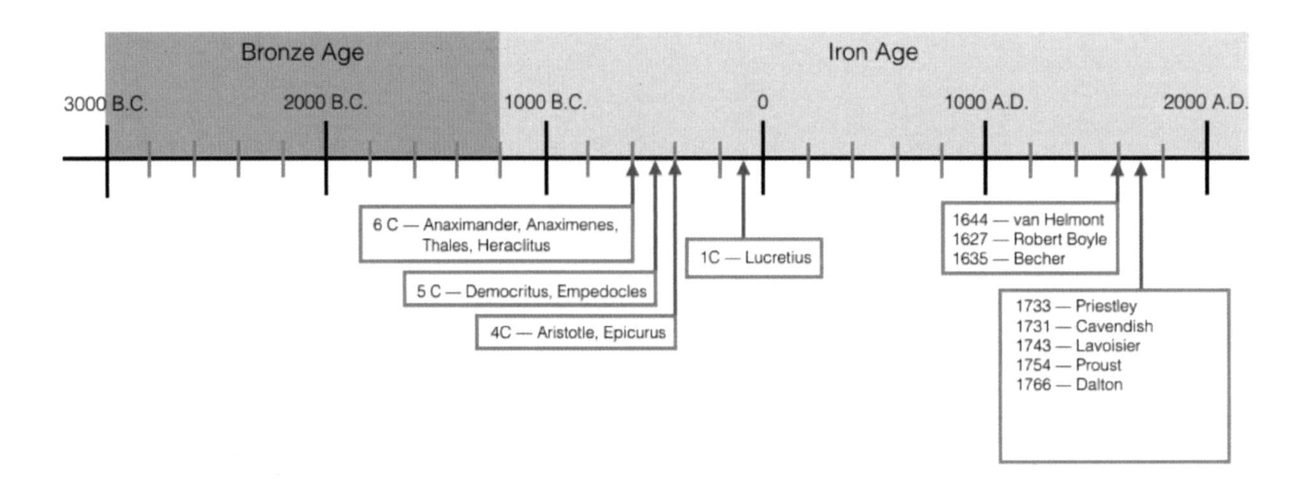

Additional Resource: Dalton had to use his imagination to envision the structure of atoms because it wasn't until the 1980s that chemists had the technology to see individual atoms. You can see them yourself if you have time to watch the video "How Can You See an Atom?" (https://qrs.ly/ybcqicx, 5:16 min).

LESSON 20

Today's reading assignment begins in the middle of the chapter you were reading during the last lesson. It starts with the line: "*That was an enormous leap forward for chemistry...*" Do you remember what the leap forward was? Take a minute to recall what you learned in the last lesson before getting started today.

Read: *The Mystery of the Periodic Table.* Chapter 13, "Mr. Dalton and His Atoms," pages 87-92. Place your bookmark at the beginning of chapter 14.

Notebook: Write what you have learned in your science notebook.

Activity: Include Dalton's symbol chart in your science notebook. You can either draw the symbols shown on page 84 of the text or draw the table shown below from Dalton's book. Alternatively, you can print Dalton's symbol chart from Wikimedia Commons (https://qrs.ly/mrcqibo) and tape or glue it into your notebook.

Additional Resource: If you are interested in learning more about John Dalton, I encourage you to spend some of your free time reading chapter 7 of the book *Crucibles* by Bernard Jaffe.

For Discussion: This chapter began by listing Dalton's "problems," but by the end of his life, his reputation was so great that 40,000 people paid their respects at his funeral. Do you think Dalton let his various problems hold him back?

Various atoms and molecules as depicted in John Dalton's *A New System of Chemical Philosophy* (1808).
Source: haade - En.wiki, [Public Domain], via Wikimedia Commons

LESSON 21—Experiment

Antoine Lavoisier's Law of Conservation of Mass and John Dalton's proposal that during chemical reactions atoms are neither created nor destroyed, but instead rearranged, form the basis of how we study chemical processes. Today, scientists balance their chemical equations to account for each atom.

An unbalanced chemical equation only tells which compounds react with each other and what the product is.

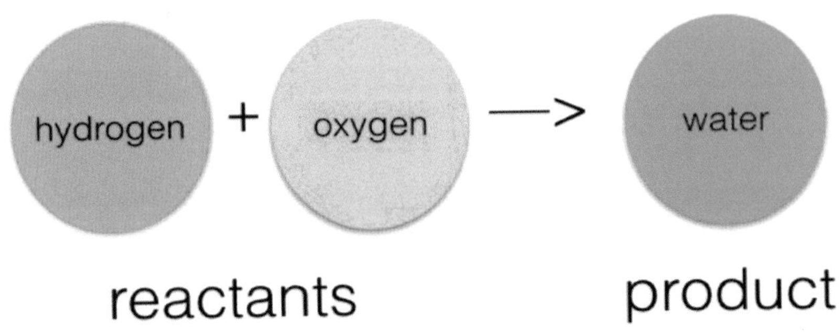

You have already learned that the substances you start with are shown on the left and are called reactants, and the substances produced by the reaction are recorded on the right and are called products. Additionally, the '+' on the left means 'reacts with' and the arrow means 'produces.' So, the equation above tells you simply that hydrogen reacts with oxygen to produce water.

A balanced equation also indicates the number of each type of atom that goes into the reaction and verifies that the same number of atoms are in the substances produced. For example, in the following image, you'll see that two molecules of hydrogen (four hydrogen atoms total) react with one molecule of oxygen (two oxygen atoms total) to produce two molecules of water (four hydrogen atoms and two oxygen atoms total). You can see that it is balanced because there are four hydrogen atoms on each side and two oxygen atoms on each side. Therefore, the same collection of atoms is present; they have only been rearranged.

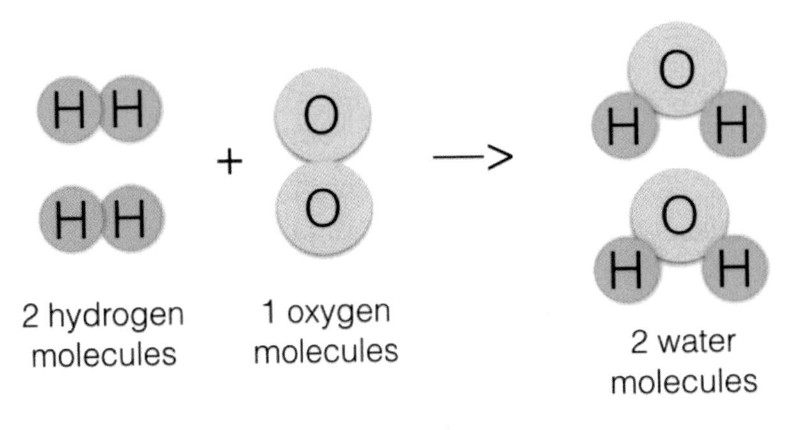

There is a more straightforward way to write the above equation, but first, you must recognize the difference between an atom and a molecule. An atom is a single unit of an element, such as one atom of oxygen. If two or more atoms of the same kind or different kinds bond together, it is called a molecule. In the image above, you see that two oxygen atoms have paired up to create a single molecule of oxygen. We count two atoms of oxygen in that one molecule of oxygen and write O_2. The subscript number tells you how many atoms of oxygen are present in that single molecule of oxygen.

There are two molecules of hydrogen, so that is handled a bit differently. A single molecule of hydrogen is written H_2; however, because there are two of those molecules, it is written $2H_2$. The number in front of the formula is called a *coefficient*. Did you notice that there is not a number one in front of O_2 to indicate that there is only one molecule of oxygen? The number one is assumed and usually omitted.

The balanced chemical equation for water is read, two hydrogen molecules react with one oxygen molecule to produce two water molecules.

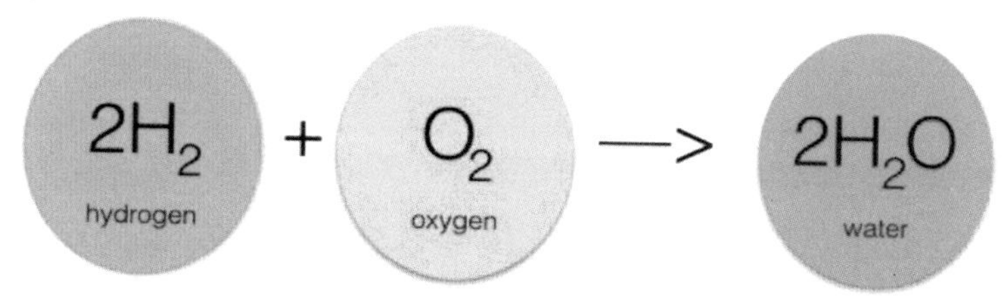

Take a few minutes to complete the activity below to secure this idea in your mind.

Activity: Observe chemical compositions before and after a chemical reaction.

Supplies Needed—

- Print out of the Chemical Composition Worksheet found at https://qrs.ly/1pcqhz9. The worksheet is in color, but you can print it as black and white if you prefer.
- Scissors
- Glue

Procedure— Follow the instructions on the worksheets.

Optional Activity: If you would like to try balancing chemical equations that are not already balanced, use the interactive "Balancing Chemical Equations" (https://qrs.ly/2ycqic3).

For Discussion: How do you think balancing an equation can help a chemist?

LESSON 22

Have you ever been given laughing gas before a procedure at the dentist's office? The dentist can use it to help calm a person who is nervous about the dental work to be done. Today you will learn about the man who first discovered laughing gas.

Sir Humphry Davy, 1830 engraving based on the painting by Sir Thomas Lawrence
Source: By unknown, [Public Domain], via Wikimedia Commons

Read: *The Mystery of the Periodic Table.* Chapter 14, "The Shocking Mr. Davy," pages 93-97. Place your bookmark at the beginning of chapter 15.

"Full of mischief, with a penchant for explosions, Davy was a born chemist."

— T. K. KENYON, *SCIENCE AND CELEBRITY: HUMPHRY DAVY'S RISING STAR*

Notebook: 1) Write what you have learned in your science notebook. 2) Add Humphry Davy to your timeline.

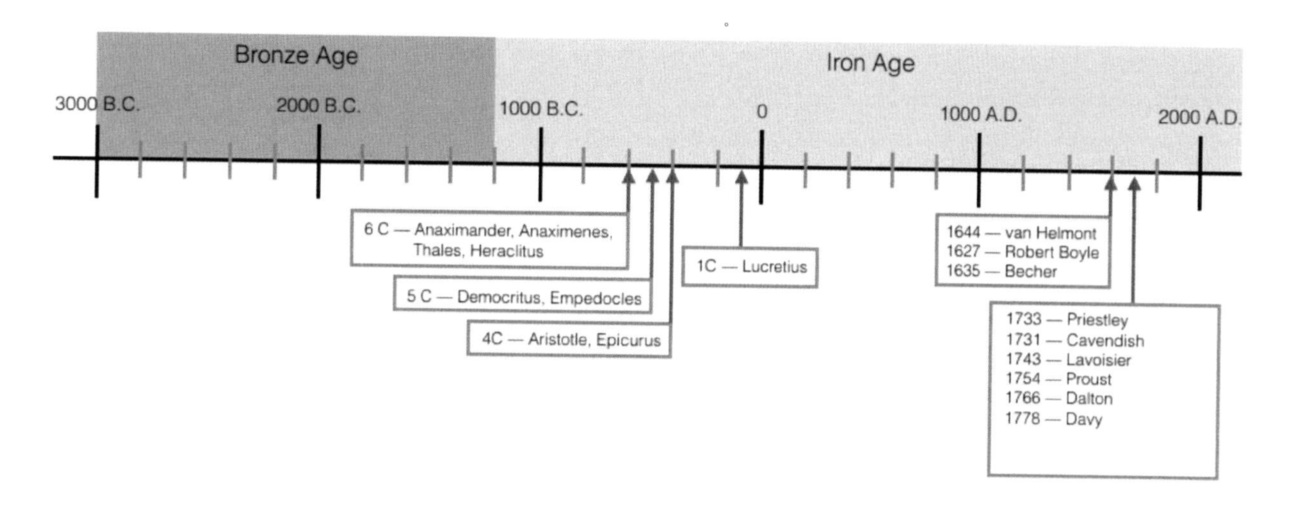

3) Look up the elements mentioned in this chapter. Mark or color the ones you haven't already noted on your Periodic Table.

1. Potassium (P) 19
2. Sodium (Na) 11
3. Magnesium (Mg) 12
4. Calcium (Ca) 20
5. Strontium (Sr) 38
6. Barium (Ba) 56
7. Chlorine (Cl) 17

For Discussion: Nitrous oxide, or laughing gas, is also used as a food additive. It is used in aerosol whipped cream canisters and cooking sprays. It's also the air used to fill packages of potato chips and other snack foods. If you are a car racing fan, you might have heard it referred to as "nitrous." How do you suppose people think of these unusual ways to use a single molecule?

LESSON 23

Recall what you learned about John Dalton in lessons 19 and 20. Specifically, remember what the Law of Definite Proportions states. Today you will read about the way Guy-Lussac built on Dalton's idea.

Gay-Lussac

Source: François Séraphin Delpech, [Public Domain], via Wikimedia Commons

Read: *The Mystery of the Periodic Table.* Chapter 15, "Guy-Lussac and Avogadro to the Rescue," pages 98-104, put your bookmark at the top of page 105.

Please Note—You may need to take this section slowly, make mental pictures as you read, and study the author's illustrations. There is no reason to rush, and taking a few extra minutes to make sure you understand what is explained, will help you in the end.

Notebook: 1) Write what you have learned in your science notebook. 2) Add Guy-Lussac to your timeline.

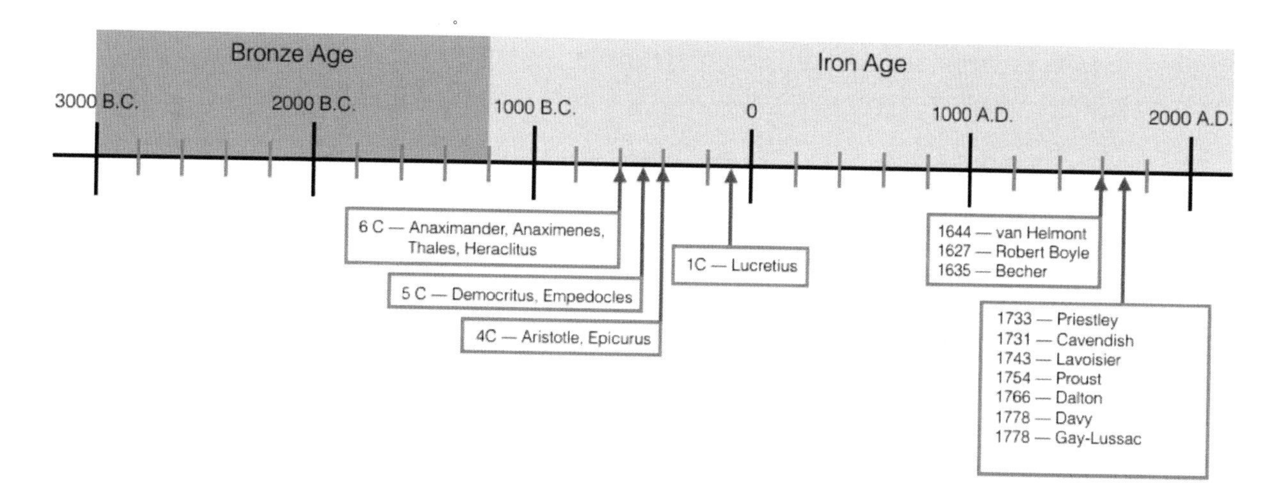

Further Study:

In 1804, Joseph Gay-Lussac made a hot-air balloon ascent with Jean-Baptiste Biot. They reached a height of 6.4 kilometers so that they could study the earth's atmosphere. They wanted to collect air samples from different altitudes to analyze the differences in temperature, pressure, and humidity. The following year he told of his findings that the earth's atmosphere does not change in composition with increasing altitude (and therefore decreasing pressure). In 1808 Gay-Lussac announced another significant discovery: Gay-Lussac's Law. Today we use Gay-Lussac's law in determining pressure and temperature differences of gas samples.

Gay-Lussac and Biot ascend in a hot air balloon, 1804. Illustration from the late 19th century.
Source: Trialsanderrors, [Public Domain], via Wikimedia Commons

LESSON 24—Experiment

"Davy's 1808 discoveries [of five elements: barium, calcium, boron, strontium, and magnesium] depended on his use of and research into the burgeoning field of electrochemistry, the study of electricity's effect on chemical reactions. As a young researcher at the Bristol Pneumatic Institute, Davy had caught the fever of excitement over Count Alessandro Volta's 1800 paper describing what came to be known as the voltaic pile, a sandwich of a damp cardboard disk between two metal disks that generated a weak but continuous charge. Young Davy immediately began to study and experiment with voltaic piles, making batteries out of them, and using the electrical charges to separate elements from their compounds. Davy had contributed to the field by discovering that electricity itself was caused by chemistry."[4]

Activity: Build your own Voltaic Pile.

Although Volta and Davy used silver and zinc, it is more feasible – and inexpensive – to use copper and zinc for the metal disks. Even though pennies are no longer made of copper, their copper coating makes them an excellent choice for copper disks, and zinc disks can be obtained by purchasing galvanized electrical boxes and punching out the holes.

Supplies Needed—

- Galvanized electrical box
- 17 pennies
- Thick card stock
- Salt water
- Paper towel
- A voltmeter or multimeter
- Scissors
- 4 wooden dowel rods or skewers
- 2" chunk of modeling clay
- 1 LED diode
- Connecting wires or alligator test leads

Please Note—If it would help you to see images or a video of how to build your Voltaic Pile, please visit Arbor Scientific (https://qrs.ly/yicqic6, 6:54 min.).

[4] Kenyon, T. K. "Science and Celebrity: Humphry Davy's Rising Star." *Chemical Heritage Foundation.* N.p., 27 Mar. 2017. Web. 03 June 2017.

Procedure—

1. Punch out the holes from the electrical box. You will use the disks.
2. Cut circles from the card stock that are the size of a penny.
3. Soak the cardstock circles in a cup of saltwater and then set them on a paper towel so they are not dripping when you are ready to use them.
4. To make a single cell, place a cardstock circle soaked in saltwater on top of a zinc disk and then place a penny on top of the card stock.
5. Touch the positive probe of a voltmeter to the copper and the negative probe to the zinc. You will find that the electric potential difference, or voltage, of this simple electrochemical cell will likely be between 0.60 V and 0.80 V.
6. Make another cell and stack it on top of the other so that you have two cells in series. You should find that the resulting electric potential difference is between 1.20 V and 1.60 V.
7. Set your cells on a piece of modeling clay and poke the wooden dowels into the clay to keep the growing stack of cells from falling over.
8. Continue adding cells on top of your pile, testing the voltage with each addition.
9. Connect the negative lead of the LED diode (the shorter of the two leads exiting the bottom of the bulb) to one of your alligator test leads and then to the negative (zinc - bottom of the pile) terminal of your voltaic pile. Just push it up next to the bottom zinc disk.
10. Connect the positive lead of the LED diode (the longer of the two leads exiting the bottom of the bulb) to one of your alligator test leads and then to the positive (copper - top of the pile) terminal of your voltaic pile. Just touch it to the top of the pile.
11. Your LED bulb should light up.

This activity was adapted from Arbor Scientific.

Notebook: Record the steps you took and what you learned from this experiment in your science notebook. Include drawings, if you would like.

For Discussion: Thinking he could make a dramatic impact by using more than 800 Voltaic Piles connected, Davy built the largest battery the world had ever seen. He joined the tips of two carbon filaments to it. When he brought the two carbon-filament tips together, continuously flowing electricity from the large battery caused an incredibly bright spark. The audience, watching Davy's experiment, was stunned by this demonstration.[5]

---

[5] Bos, Carole. "Discovering Electricity - Humphry Davy and the Arc Light." *AwesomeStories.com*. N.p., n.d. Web. 03 June 2017.

LESSON 25

Today's reading assignment begins in the middle of the chapter. Do you remember what you were learning the last time you read from the text? Take a minute to narrate to yourself before getting started today.

The story picks up with Lorenzo Romano Amedeo Carlo Avogadro, hailed as a founder of the atomic-molecular theory.

Amedeo Avogadro
Source: Anton, via Wikimedia Commons

Read: *The Mystery of the Periodic Table.* Chapter 15, "Guy-Lussac and Avogadro to the Rescue," pages 102-108. Place your bookmark at the beginning of chapter 16.

Please Note—Again today, you may need to read this section slowly. Make mental pictures as you read and study the illustrations provided by the author.

Notebook: 1) Write what you have learned in your science notebook. 2) Add Amedeo Avogadro to your timeline.

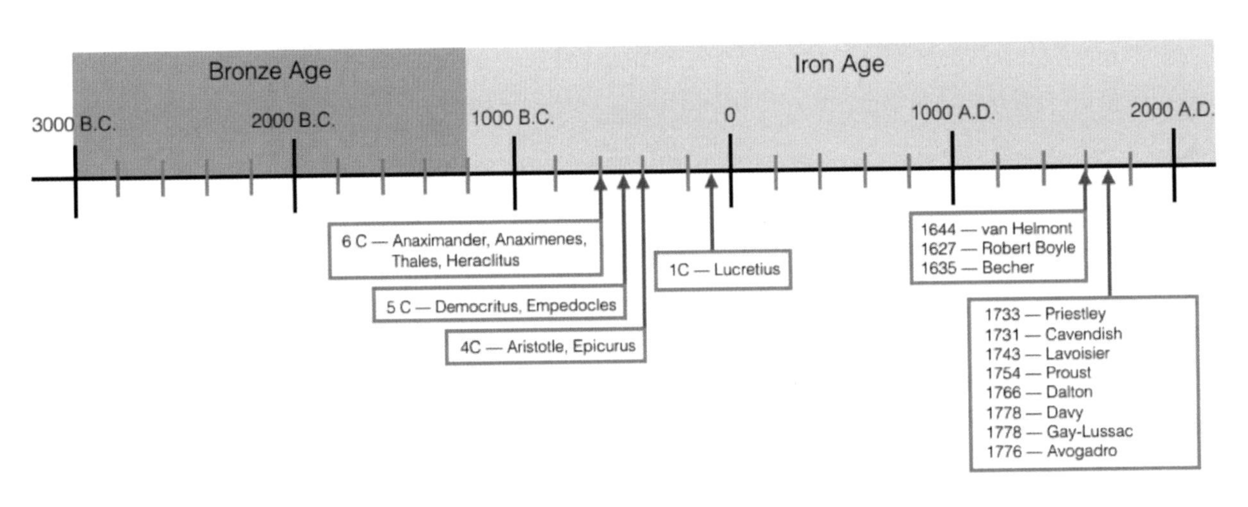

Additional Resource: If you are interested in learning more about Amedeo Avogadro, I encourage you to spend some of your free time reading chapter 9 of the book *Crucibles* by Bernard Jaffe.

LESSON 26

Read: *The Mystery of the Periodic Table.* Chapter 16, "Things Fall into Place: Triads and Octaves," pages 109-116. Place your bookmark at the beginning of chapter 17.

Please Note—Have your science notebook open to the periodic table as you read today's assignment. You will need to refer to it often.

Notebook: Write what you have learned in your science notebook.

LESSON 27—Experiment

In Lesson 18, you learned some of the ways you can observe that a chemical reaction has taken place without performing a chemical analysis of the products. You have already seen chemical reactions that produce a color change, the formation of gas, and the formation of a precipitate. Today you will observe chemical reactions that involve an increase or decrease in temperature.

You know that during chemical reactions, atoms are neither created nor destroyed but rather rearranged. Still, you may not realize how much energy is required to break the strong bonds between the atoms of the reactants and how much energy is released when the new bonds are created to make the products.

Consider, once again, the chemical reaction that takes place when you combine baking soda and vinegar. A greater amount of energy was required to break the bonds (indicated by the large red arrows below) than was released when the new bonds were formed (indicated by the small red arrows.)

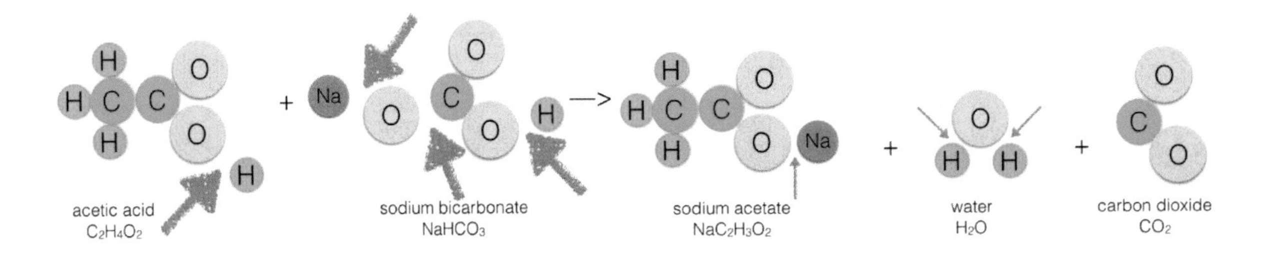

In this case, more energy went into breaking the bonds than was released when the bonds were formed, so this is an endothermic reaction. Had you taken the temperature of your product when you completed this experiment, you would have noticed a decrease in temperature.

An exothermic reaction occurs when the energy used to break the bonds in the reactants is less than the energy released when new bonds are made in the products. This extra energy is given off as heat, and you would notice an increase in temperature.

Activity: Observe an endothermic and exothermic reaction.

Supplies Needed—

- Rubber gloves
- Safety glasses

- Graduated cylinder or use a long-necked bottle (such as an empty, and clean, individual size soda bottle)
- 50 ml (1.5 oz) hydrogen peroxide (6% solution)
- 1 tablespoon dry yeast or ¼ teaspoon potassium iodide (Both work equally well. It just depends on whether you would like to use a product that is a common household item or you would like to use a product that seems more *scientific*.)
- 1 tablespoon liquid dish soap
- 8 drops food coloring (optional)
- Cake pan
- Room temperature water
- Small cup
- Funnel
- Candy or meat thermometer
- Beaker, 250 ml or a glass cup
- Ammonium nitrate. Alternatively, you can cut open a cold pack bag to remove the packet of little white balls. It's full of ammonium nitrate.

Procedure—

PART 1

1. Put on safety goggles and gloves.
2. Pour 50 ml hydrogen peroxide into the graduated cylinder.
3. Record the temperature of the hydrogen peroxide in your science notebook.
4. Add approximately a tablespoon of liquid dish soap to the graduated cylinder. Swirl the mixture to combine it by holding the top of the cylinder and rotating the bottom.
5. Add approximately 8 drops of food coloring to the cylinder. Swirl it again.
6. Set the graduated cylinder in the middle of a cake pan.
7. If you are using yeast rather than potassium iodide, combine 3 tablespoons of room temperature water with 1 tablespoon of dry yeast in a separate cup. Stir together for about 30 seconds.
8. Record the temperature of the yeast mixture in your science notebook.
9. Using a funnel, pour the yeast mixture into the cylinder or add ¼ teaspoon potassium iodide to it.
10. Observe the chemical reaction.
11. Hold a thermometer at the top of the cylinder to determine the temperature of the product. Record the temperature of the product in your science notebook.

This activity was adapted from Science Made Simple.

Notebook: 1) Record the steps you took and what you learned from this experiment in your science notebook. 2) Did this experiment demonstrate an endothermic reaction, making the solution colder, or an exothermic reaction, causing the solution to become warmer? Include drawings, if you would like.

PART 2

1. Put on safety goggles and gloves.
2. Fill a beaker with 100 ml (3-4 oz) of water.
3. Record the temperature of the water in your science notebook.
4. Add approximately 3 spoonfuls of ammonium nitrate and stir.
5. Record the temperature of the ammonium nitrate in your science notebook.

When ammonium nitrate is dissolved in water, it breaks down into its ions: ammonium and nitrate. Neither the water nor ammonium nitrate is changed in any other way other than the ammonium nitrate is dissolved. However, when aqueous or dissolved ammonium nitrate is heated, the solution breaks down to release nitrous oxide or laughing gas.

Please Note—You can wash the solutions of both experiments down the sink when you clean up.

This activity was adapted from Periodic Videos.

Notebook: Record the steps you took and what you learned from the experiment in your science notebook. Did this experiment demonstrate an endothermic reaction, making the solution colder, or an exothermic reaction, causing the solution to become warmer? Include drawings, if you would like.

Optional Activity: If you would like to use some of your free time to try a more involved exothermic reaction, follow the instruction provided by The Sci Guys: Science at Home - SE1 - EP7: Hot Ice - Exothermic Reactions and Supercooled solutions (https://qrs.ly/ercqicf, 6:17 min.) This activity will take several hours but produces very impressive results.

For Discussion: Use what you know about Latin roots to explain the meaning of the words endothermic and exothermic.

LESSON 28

Dmitri Mendeleev
Source: Journal of Chemical Education

In lesson 26, you read about John Newland, who read a paper before the English Chemical Society proposing a way to arrange the elements. They laughed at him. Only three years later, Dmitri Mendeleev proposed an almost identical idea. Would they respond to him in kind?

Read: *The Mystery of the Periodic Table.* Chapter 17, "The Mystery Solved," pages 117-124, put your bookmark at the top of page 125.

Please Note—Have your science notebook open to the periodic table as you read today's assignment. You will need to refer to it often.

Notebook: 1) Write what you have learned in your science notebook. 2) Add Dmitri Mendeleev to your timeline.

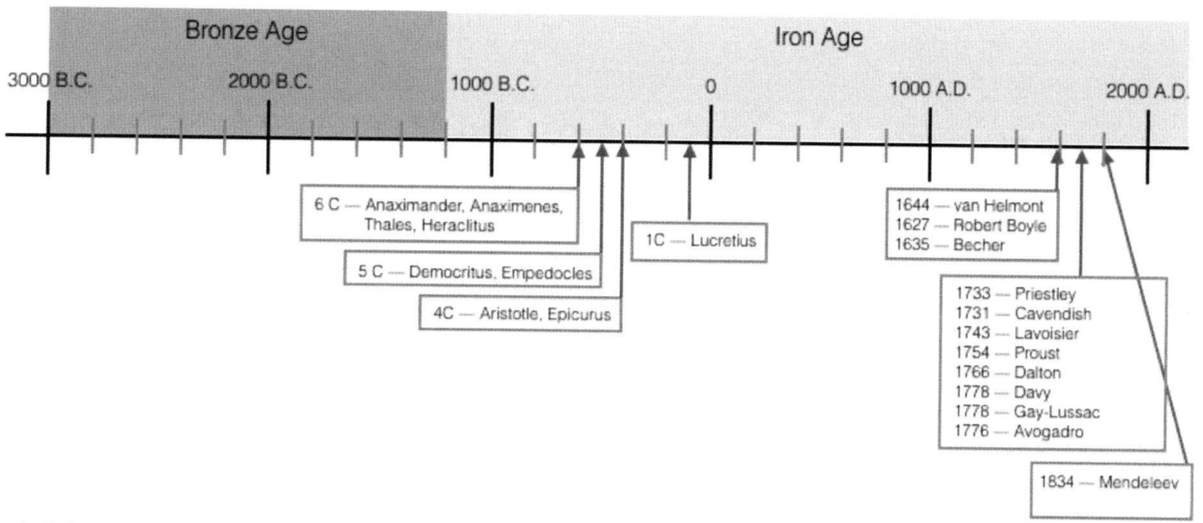

Additional Resource: If you are interested in learning more about Dmitri Mendeleev, I encourage you to spend some of your free time reading chapter 11 of the book *Crucibles* by Bernard Jaffe.

For Discussion: On page 119, you read: "*So the Periodic Table merely means that, for some mysterious reason, the elements are not just thrown together by nature in a jumble, but show very regular, astoundingly regular, patterns—if we are clever enough to uncover that underlying order.*"

LESSON 29

Today you will continue to learn about Mendeleev and his contribution to the list of known elements.

Read: *The Mystery of the Periodic Table.* Chapter 17, "The Mystery Solved," pages 125-128. Place your bookmark at the beginning of chapter 18.

Notebook: 1) Write what you have learned in your science notebook. 2) Look up the elements mentioned in this chapter. Mark or color the ones you haven't already noted on your Periodic Table.

1. Boron (B) 5
2. Scandium (Sc) 21
3. Aluminum (Al) 13
4. Calcium (Ca) 20
5. Zinc (Zn) 30
6. Arsenic (As) 33
7. Gallium (Ga) 31
8. Germanium (Ge) 32
9. Indium (In) 49
10. Cadmium (Cd) 48
11. Tin (Sn) 50
12. Uranium (U) 92
13. Gold (Au) 79
14. Mendelevium (Md) 101

For Discussion: At the same time, Mendeleev conceived of his Periodic Law, De Chancourtois in France, Stretcher in Germany, Newlands in England, and Cooke in America had noticed the same similarities among the properties of the known elements.[6] Charlotte Mason tells us: "*We may believe also ... that a revelation is still going on of things not hitherto made known to men. Great secrets of nature, for example, would seem to be imparted to minds already prepared to receive them, as, for example, that of the 'ions' or 'electrons' of which that we call matter is said to consist. For this sort of knowledge also is of God, and is, I believe, a matter of revelation, given as the world is prepared to receive it.*" (Ourselves, p. 86-87) As you have read the history of chemistry, you have learned about the men given credit for each new revelation. However, there have almost always been two or three other men who fought for that recognition. Do you think it is appropriate for only one person to receive all the praise?

---

[6] Jaffe, Bernard. *Crucibles the Story of Chemistry from Ancient Alchemy to Nuclear Fission.* New York: Dover Publications, 1976. 158. Print.

LESSON 30—Experiment

In the last experiment, you learned that a chemical reaction requires energy to break the strong bonds between the atoms of the reactants. Today, you will learn that adding more energy by adding heat makes the reactants move faster, making more of them collide and collide harder, increasing the reaction rate.

Activity: Test whether temperature affects the rate of a chemical reaction.

Supplies Needed—

- Rubber gloves
- Safety glasses
- 3 glow sticks
- Masking tape
- Pen
- Calcium chloride, granular, **an irritant** OR Road Salt (Read the label to make sure it is pure calcium chloride; sometimes other chemicals are added to help melt snow.)
- Baking soda
- Water
- Graduated cylinder or measuring cup
- Digital scale or ½ teaspoon measuring spoon
- 2 wide (9 oz) clear plastic cups
- 4 small clear plastic cups
- 2 plastic deli-style condiment containers
- Hot water (40–50 °C)
- Ice water (0–5 °C)
- A helper

Procedure—

PART 1

1. Fill a cup with hot water and another cup with ice water.
2. Place one glow stick in the hot water and another in the ice water and leave them for five minutes. Be careful not to "start" the glow sticks.
3. Leave the other glow stick on the table, so it remains at room temperature.
4. Remove the glow sticks from both the hot and cold water.
5. Bend them until you hear a popping sound. This will cause them to glow.
6. Observe the brightness of each stick.
    a. Does one of them glow brighter than the other?
    b. Can you tell whether the chemical reaction is happening faster or slower in each glow stick?
    c. Which stick will stop glowing the soonest?
    d. Some people place glow sticks in the freezer to make them last longer. Why do you think this works?
    e. Why do you think that starting with warmer reactants increases the rate of other chemical reactions?

PART 2

1. Use masking tape and a pen to label two small clear plastic cups "baking soda solution" and two small plastic cups "calcium chloride solution."
2. Make a Baking Soda Solution
    a. Use a graduated cylinder to add 20 ml (about 4 teaspoons) of water to one of the baking soda solution cups.
    b. Add 2 g (about ½ teaspoon) of baking soda to the water in its labeled cup.
    c. Swirl until as much of the baking soda dissolves as possible. (There may be some undissolved baking soda in the bottom of the cup.)
    d. Pour half of your baking soda solution into the other baking soda solution cup.
3. Make the Calcium Chloride Solution
    a. Use a graduated cylinder to add 20 ml (about 4 teaspoons) of water to one of the calcium chloride solution cups.
    b. Add 2 g (about ½ teaspoon) of calcium chloride to the water in its labeled cup.
    c. Swirl until the calcium chloride dissolves.
    d. Pour half of your calcium chloride solution into the other calcium chloride solution cup.
4. Heat and Cool the Solutions
    a. Pour hot water into one plastic container and cold water into the other until each is about ¼ filled. The water should not be very deep. These are your hot and cold water baths.
    b. Place and hold one cup of baking soda solution and one cup of calcium chloride solution in the hot water. Gently swirl the cups in the water for about 30 seconds to heat the solutions.
    c. At the same time, you and your partner should combine the two warm solutions. Also, combine the two cold solutions.
5. Combine the Solutions
    a. At the same time, you and your partner should combine the two warm solutions with each other, and the two cold solutions with each other.

Notebook: 1) Notebook: Record the steps you took and what you learned from this experiment in your science notebook. 2) Did the temperature of the reactants affect the rate of the chemical reaction? Include drawings, if you would like.

These activities were adapted from Middle School Chemistry.

LESSON 31

In Lesson 22, you read that Sir Humphry Davy suspected elements themselves must somehow be electrically positive and negative. "*It would take over one hundred years to see how absolutely marvelous Davy's suspicion was.*" (p. 97) You have arrived at that date.

Ernest Rutherford

Read: *The Mystery of the Periodic Table.* Chapter 18, "The Mystery Continues," pages 129-136, put your bookmark on page 136 before "*But why do the elements line up vertically?*"

Notebook: 1) Write what you have learned in your science notebook. 2) Include a drawing of the atom shown at the bottom of page 134. Label all of the parts. You must know the parts of the atom before completing the next lesson. 3) Add Ernest Rutherford to your timeline.

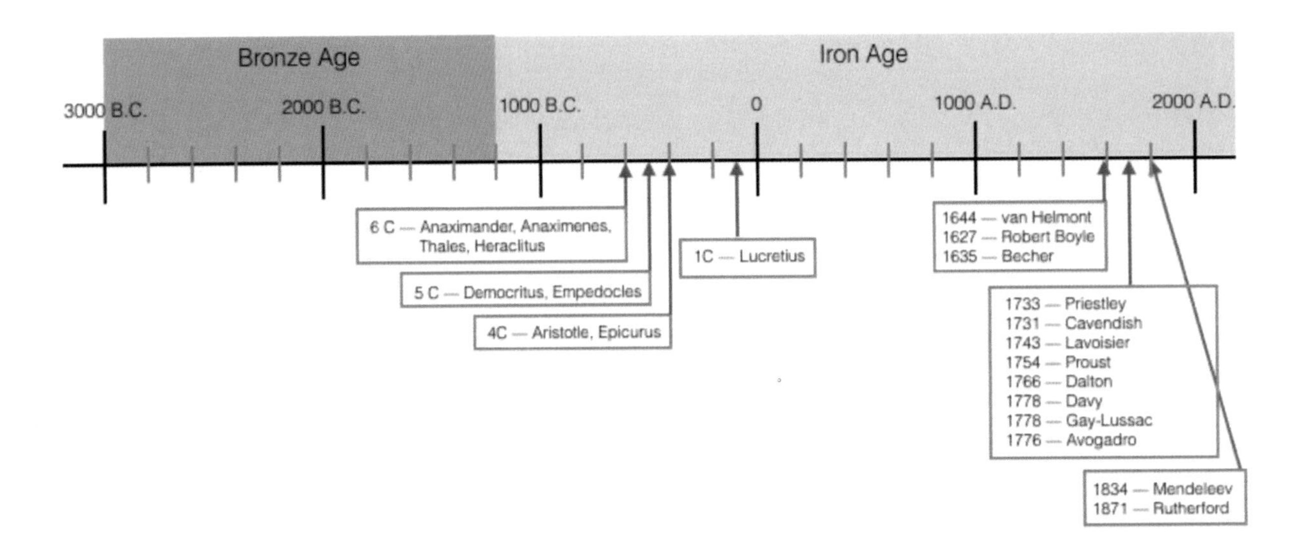

If you would like to, you can draw the following diagram depicting the conclusions of Rutherford's gold foil experiment in your science notebook.

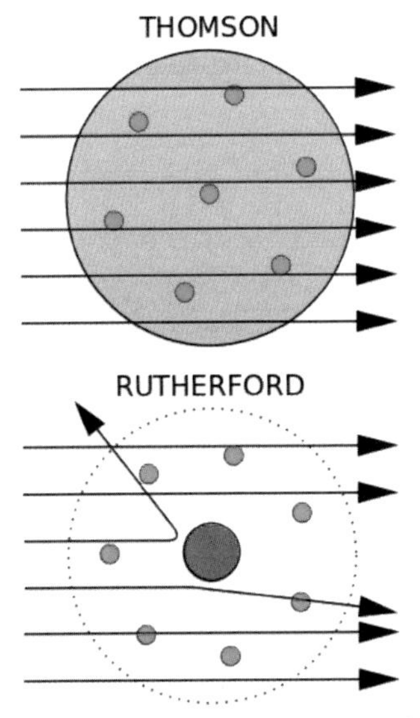

Conclusions of Rutherford's gold foil experiment.
Source: Kurzon, [Public Domain], via Wikimedia Commons

Additional Resource: If you are interested in learning more about Ernest Rutherford, I encourage you to spend some of your free time reading chapter 14 of the book *Crucibles* by Bernard Jaffe.

Also, if you would like to view a reproduction of Rutherford's gold foil experiment, you can watch the video "Rutherford Gold Foil Experiment - Backstage Science." (https://qrs.ly/k4cqich, 4:05 min.)

LESSON 32

Do you remember what you learned the last time you read from the text? Take a minute to narrate to yourself before getting started today. For example, do you recall the parts of an atom? It might benefit you to look at the drawing of an atom you included in your science notebook during the last lesson.

Read: *The Mystery of the Periodic Table.* Chapter 18, "The Mystery Continues," pages 136-145, put your bookmark at Part II on page 145.

Please Note—Read this section slowly, making mental pictures as you go, and studying each illustration provided by the author. It may also be helpful to have your science notebook open to the periodic table as you read today's assignment.

Notebook: 1) Write what you have learned in your science notebook. Be sure to include an explanation for why the elements line up vertically, why elements in Group 1A are so reactive, while elements in Group 8A are so stand-offish, and what an isotope is. 2) Reproduce some or all of the diagram on pages 138 - 139.

LESSON 33

Today you will complete your study of the history of chemistry!

Read: *The Mystery of the Periodic Table.* Chapter 18, "The Mystery Continues," pages 145-151. Place your bookmark at the beginning of chapter 19.

Please Note—Have your science notebook open to the periodic table as you read today's assignment.

Notebook: 1) Write what you have learned in your science notebook. 2) Explain the way the electron shells fill in the transition elements compared to the main elements. 3) Look up the elements mentioned in this chapter. Mark or color the ones you haven't already noted on your Periodic Table.

1. Uranium (U) 92
2. Neptunium (Np) 93
3. Meitnerium (Mt) 109
4. Ununnilium (Uun) 110 (now Darmstadtium)
5. Unununium (Uuu) 111 (now Roentgenium)
6. Ununbium (Uub) 112 (now Copernicium)
7. Ununquadium (Uuq) 114 (now Flerovium)
8. Ununhexium (Uun) 116 (now Livermorium)

Additional Resource: In the text, you read: "*Element 113? I've not heard anything yet. Have you? ...Element 115? No news here.*" You may like to read the article "Four new elements officially added to the periodic table" (https://qrs.ly/hmcqicp) for an update.

You might also like to complete your chemistry study with "The Periodic Table Song" (https://qrs.ly/rrcqict). If nothing else, you may want to know how the names of some of those elements are pronounced.

About the Author

Nicole Williams learned about Charlotte Mason a few years after she began homeschooling and the same year, she added three additional students to her schoolroom! It was a trial by fire that resulted in a refinement of Charlotte Mason's methods and philosophy in her home. More than a decade later, she has written a living science curriculum, teaches at conferences, and co-hosts the Charlotte Mason podcast A Delectable Education. Nicole enjoys working in her garden, collecting living books, hiking, reading, and listening to podcasts.

Other Titles in This Series

Form 2 (Grades 2-6)

Botany. The First Book of Plants by Alice Dickinson
Chemistry. Matter, Molecules, and Atoms by Bertha Morris Parker
Physics - Magnets. Magnets by Rocco V. Feravolo
Physics - Waves. The First Book of Sound by David Knight
Physics - Energy. The First Book of Electricity by Sam and Beryl Epstein
Engineering & Technology. The First Book of Machines by Walter Buehr
Astronomy. Find the Constellations by H.A. Rey
Weather. Rain, Hail, Sleet & Snow by Nancy Larrick
Geology. The First Book of the Earth by O. Irene Sevrey

Form 3-4 (Grades 7-9)

Biology. Men, Microscopes, and Living Things by Katherine B. Shippen
Botany. First Studies of Plant Life by George Francis Atkinson
Chemistry. The Mystery of the Periodic Table by Benjamin Wiker
Physics. Secrets of the Universe by Paul Fleisher
Engineering & Technology. Electronics for Kids by Øyvind Nydal Dahl
Astronomy. The Planets by Dava Sobel
Weather. Look at the Sky and Tell the Weather by Eric Sloane
Geology. Rocks, Rivers and the Changing Earth: A First Book About Geology by Herman and Nina Schneider

High School (Grades 9-12)

Biology, Anatomy part 1. The Body: A Guide for Occupants by Bill Bryson
Biology, Anatomy part 2. The Body: A Guide for Occupants by Bill Bryson
Biology, Ecology.
Biology, Origins.
Chemistry, part 1. Wonders of Chemistry by A. Frederick Collins
Chemistry, part 2. Wonders of Chemistry by A. Frederick Collins
Chemistry, part 3. Wonders of Chemistry by A. Frederick Collins
Physics, part 1. For the Love of Physics by Walter Lewin
Physics, part 2. For the Love of Physics by Walter Lewin
Physics, part 3/Astrophysics. For the Love of Physics by Walter Lewin
Geology. Aerial Geography by Mary Caperton Morton
Weather.

Made in the USA
Las Vegas, NV
07 January 2025